Contributions to Higher Engineering Education

Maria M. Nascimento · Gustavo R. Alves
Eva Virgínia Araújo Morais
Editors

Contributions to Higher Engineering Education

 Springer

Editors
Maria M. Nascimento
Department of Mathematics, School
  of Science and Technology
University of Trás-os-Montes e Alto Douro
Vila Real
Portugal

Eva Virgínia Araújo Morais
Departamento de Matemática
University of Trás-os-Montes e Alto Douro
Vila Real
Portugal

Gustavo R. Alves
Instituto Superior de Engenharia do Porto
Porto
Portugal

ISBN 978-981-10-8916-9   ISBN 978-981-10-8917-6   (eBook)
https://doi.org/10.1007/978-981-10-8917-6

Library of Congress Control Number: 2018937320

Printed on acid-free paper

This Springer imprint is published by the registered company Springer Nature Singapore Pte Ltd.
part of Springer Nature
The registered company address is: 152 Beach Road, #21-01/04 Gateway East, Singapore 189721,
Singapore

Preface

Education is the most powerful weapon, which you can use to change the world.

Nelson Mandela[1]

The world is full of new challenges and opportunities. If we look around, energy, climate change, aging population, economic development, and a lot of other issues, we see engineers playing a key role in implementing new solutions. The new approaches required are extremely valuable all through every engineering field, as they enable practical and creative actions in the face of a volatility, uncertainty, complexity, and ambiguity world. Therefore, the engineers' education is of a paramount importance.

Engineering education is about facilitating the learning of scientific and technical knowledge, as well as learning the principles of the professional practice of engineering. Nowadays, such professional practice requires other skills to train a twenty-first century engineer, known as soft skills, such as teamwork, creativity, and ability to communicate.

In Portugal, the stakeholders of the engineering education *process* got together to think how to improve their own work to generate the future professionals of the several engineering domains. That effort led to the foundation of the Portuguese Society for Engineering Education (SPEE). Afterward, the Society's international conferences began, and on its second meeting (CISPEE 2016) writing a book with the more relevant contributions was proposed. This initiative was encouraged by SPEE to promote engineering education. The six chapters of this book will offer the opportunity to spread some of the works in this field in Portugal.

The first chapter of this book, entitled "International Cooperation for Remote Laboratory Use," corresponds to the best paper presented at CISPEE 2016, held at the University of Trás-os-Montes e Alto Douro, Vila Real, Portugal, in October 2016, and it presents a very interesting cooperative work between several

---

[1] Source: "Lighting your way to a better future: Speech delivered by Mr. N. R. Mandela at launch of Mindset Network", July 16, 2003.

universities in Europe and in Latin America. It describes the VISIR+ Project which is an example of cooperative work in the field of engineering education.

The second chapter is about "Mature Learners' Participation in Higher Education and Flexible Learning Pathways: Lessons Learned from an Exploratory Experimental Research." It analyzes the students' ways of learning in higher education institutions.

Following the "The Flow of Knowledge and Level of Satisfaction in Engineering Courses Based on Students' Perceptions," the third chapter gives us a glimpse of some important aspects in engineering training and provides elements to eventually improve it.

The fourth chapter, entitled "Innovative Methodologies to Teach Materials and Manufacturing Processes in Mechanical Engineering," serves as a testimony of a different work approach implemented in engineering in higher education training.

As an example of the work integration between a higher education institution and a company, "'Learning by Doing' Integrated Project Design in a Master Program on Product and Industrial Design" is the fifth chapter and depicts situations where the students were involved in a real project, mimicking the situation where they were already operating in the job market.

Finally, the sixth chapter presents "The Views of Engineering Students on Creativity," another hint to some emphasized skills needed by the twenty-first-century engineer, and ultimately it enables us to picture its use in engineers training.

By the collaborative effort that all the authors made to bring us these best practices, I know this book will reach the global community of engineering education and that its spreading will lead to the discussion of the several challenges that will continue to emerge all over the world.

José Carlos Quadrado
President LACCEI, Past President IFEES
Past President ASIBEI

Contents

Editors and Contributors

About the Editors

Maria M. Nascimento has been an Assistant Professor at the University of Trás-Montes e Alto Douro (UTAD), Vila Real, Portugal, since 1985, where she teaches Statistics and Operations Research. She is an integrated member as a researcher of CIDTFF (Research Centre on Didactics and Technology in the Education of Trainers, CIDTFF, from University of Aveiro, and in its Lab-DCT/UTAD, the laboratory of CIDTFF at UTAD). Her main research and publication interests are teaching statistics and its attitudinal and didactical research issues, as well as the ethnomathematics research field. In addition, the critical thinking research field and in its connections to the statistical thinking in higher education are her other research interests. She is a Member of the Portuguese Engineers Society (OE) and of the Portuguese Society for Engineering Education (SPEE).

Gustavo R. Alves graduated in 1991 and obtained an M.Sc. and a Ph.D. degree in Computers and Electrical Engineering in 1995 and 1999, respectively, from the University of Porto, Portugal. He is a Professor at the Polytechnic of Porto—School of Engineering (ISEP) since 1994. He has authored or co-authored more than 200 conference and journal papers with referee process, 11 books, and co-edited a book (with Javier Garcia-Zubia, University of Deusto, Bilbao, Spain), about "Using Remote Labs in Education." He has also been involved in more than 18 national and international research projects. His research interests include engineering education, remote experimentation, and design for debug and test. He served as the Program Co-Chair of the First and Second International Conference of the Portuguese Society for Engineering Education (CISPEE 2013 and CISPEE 2016), and of the 3rd Experiment@ International Conference (exp.at'15), as General Chair of the 11th Remote Engineering and Virtual Instrumentation (REV 2014) conference and of the 3rd Technological Ecosystems for Enhancing Multiculturality (TEEM 2015) Conference, and also as a Program Committee Member of several international conferences. He is currently Associate Editor for the IEEE

Transactions on Learning Technologies and has also served as Guest Editor for the *International Journal of Online Engineering* (iJOE), the *International Journal of Engineering Pedagogy* (iJEP), the *IEEE Latin-American Learning Technologies Journal* (IEEE-RITA), and the *European Journal of Engineering Education* (EJEE).

Eva Virgínia Araújo Morais is an Assistant Professor at the University of Trás-os-Montes and Alto Douro (UTAD) since 2000. She received her Ph.D. in Mathematics Applied to Economy and Management from the Lisbon School of Economics and Management (ISEG), Technical University of Lisbon, Portugal, in 2013. She has authored and co-authored several works in her main research fields, mathematical finance and critical thinking in higher education.

Contributors

Nival N. de Almeida Brazilian Association of Engineering Education, Brasília, DF, Brazil

Gustavo R. Alves School of Engineering, Centre for Innovation in Engineering and Industrial Technology, Polytechnic of Porto, Porto, Portugal; Centro de Inovação em Engenharia e Tecnologia Industrial, Instituto Superior de Engenharia do Porto (ISEP), Porto, Portugal

João B. Alves Campus Araranguá, Federal University of Santa Catarina, Mato Alto, Araranguá, Santa Catarina, Brazil

Gaston S. de Arregui Faculty of Engineering and Exact Sciences, National University of Rosario, Rosario, Santa Fe, Argentina

Brenda Bertramo Research Institute of Rosario in Educational Sciences, National Scientific and Technical Research Council, Santa Fe, Argentina

Simone Biléssimo Campus Araranguá, Federal University of Santa Catarina, Mato Alto, Araranguá, Santa Catarina, Brazil

Daniel D. de Bona Department of Electronics, Federal Institute of Education, Science and Technology of Santa Catarina, Florianópolis, Santa Catarina, Brazil

Isabel M. Brás Pereira Centro de Inovação em Engenharia e Tecnologia Industrial, Instituto Superior de Engenharia do Porto (ISEP), Porto, Portugal

Helena Campos Departamento de Matemática da Escola de Ciências e Tecnologia da, Vila Real Universidade de Trás-os-Montes e Alto Douro—UTAD, Vila Real, Portugal

Vitor Carneiro Mestrado em Design Industrial e de Produto, Design Studio FEUP, Porto, Portugal

Manuel Castro Department of Control, Electrical and Electronics Engineering, National Distance Education University, Madrid, Spain

Paula Catarino Departamento de Matemática da Escola de Ciências e Tecnologia da, Vila Real Universidade de Trás-os-Montes e Alto Douro—UTAD, Vila Real, Portugal

Sonia B. Concari Faculty of Engineering and Exact Sciences, National University of Rosario, Rosario, Santa Fe, Argentina

Ricardo J. Costa School of Engineering, Centre for Innovation in Engineering and Industrial Technology, Polytechnic of Porto, Porto, Portugal

Elsa Dobboletta Research Institute of Rosario in Educational Sciences, National Scientific and Technical Research Council, Santa Fe, Argentina

Rogério Duarte Escola Superior de Tecnologia de Setúbal, Instituto Politécnico de Setúbal, Setúbal, Portugal

Teresa P. Duarte University of Porto, Porto, Portugal

Gabriel Díaz-Orueta Department of Control, Electrical and Electronics Engineering, National Distance Education University, Madrid, Spain

Manuel C. Felgueiras School of Engineering, Centre for Innovation in Engineering and Industrial Technology, Polytechnic of Porto, Porto, Portugal

Rubén A. Fernández Faculty of Technologies and Exact Sciences, National University of Santiago del Estero, Santiago del Estero, Argentina

Golberi Ferreira Department of Electronics, Federal Institute of Education, Science and Technology of Santa Catarina, Florianópolis, Santa Catarina, Brazil

André V. Fidalgo School of Engineering, Centre for Innovation in Engineering and Industrial Technology, Polytechnic of Porto, Porto, Portugal

Felix García-Loro Department of Control, Electrical and Electronics Engineering, National Distance Education University, Madrid, Spain

Javier García-Zubía Faculty of Engineering, University of Deusto, Bilbao, Spain

Ângela Gomes Mestrado em Design Industrial e de Produto, Design Studio FEUP, Porto, Portugal

Anabela Guedes Centro de Inovação em Engenharia e Tecnologia Industrial, Instituto Superior de Engenharia do Porto (ISEP), Porto, Portugal

Ingvar Gustavsson Department of Signal Processing, Blekinge Institute of Technology, Karlskrona, Sweden

Mario J. Gómez Faculty of Technologies and Exact Sciences, National University of Santiago del Estero, Santiago del Estero, Argentina

Romeu Hausmann Department of Electrical and Telecommunication Engineering, University of Blumenau, Blumenau, Brazil

Unai Hernández-Jayo Faculty of Engineering, University of Deusto, Bilbao, Spain

M. João Monteiro Departamento de Matemática da Escola de Ciências e Tecnologia da, Vila Real Universidade de Trás-os-Montes e Alto Douro—UTAD, Vila Real, Portugal

Wlodek J. Kulesza Department of Signal Processing, Blekinge Institute of Technology, Karlskrona, Sweden

Federico Lerro Faculty of Engineering and Exact Sciences, National University of Rosario, Rosario, Santa Fe, Argentina

Delberis A. Lima Department of Electrical Engineering, Pontifical Catholic University of Rio de Janeiro, Gávea, Rio de Janeiro, Brazil

Natércia Lima School of Engineering, Centre for Innovation in Engineering and Industrial Technology, Polytechnic of Porto, Porto, Portugal

J. Lino Alves University of Porto, Porto, Portugal

Jorge Lino Mestrado em Design Industrial e de Produto, Design Studio FEUP, Porto, Portugal

Susana Marchisio Faculty of Engineering and Exact Sciences, National University of Rosario, Rosario, Santa Fe, Argentina

A. T. Marques University of Porto, Porto, Portugal

Maria A. Marques School of Engineering, Centre for Innovation in Engineering and Industrial Technology, Polytechnic of Porto, Porto, Portugal

Claudio Merendino Faculty of Engineering and Exact Sciences, National University of Rosario, Rosario, Santa Fe, Argentina

Eva Morais Departamento de Matemática da Escola de Ciências e Tecnologia da, Vila Real Universidade de Trás-os-Montes e Alto Douro—UTAD, Vila Real, Portugal

Maria M. Nascimento Departamento de Matemática da Escola de Ciências e Tecnologia da, Vila Real Universidade de Trás-os-Montes e Alto Douro—UTAD, Vila Real, Portugal

Kristian Nilsson Department of Signal Processing, Blekinge Institute of Technology, Karlskrona, Sweden

Ângela Lacerda Nobre Escola Superior de Tecnologia de Setúbal, Instituto Politécnico de Setúbal, Setúbal, Portugal

Ana Luísa de Oliveira Pires Escola Superior de Tecnologia de Setúbal, Instituto Politécnico de Setúbal, Setúbal, Portugal

Vanderli F. de Oliveira Brazilian Association of Engineering Education, Brasília, DF, Brazil

Celina P. Leão Centro ALGORITMI, School of Engineering, University of Minho, Guimarães, Portugal

Fernando S. Pacheco Department of Electronics, Federal Institute of Education, Science and Technology of Santa Catarina, Florianópolis, Santa Catarina, Brazil

Ana M. Pavani Department of Electrical Engineering, Pontifical Catholic University of Rio de Janeiro, Gávea, Rio de Janeiro, Brazil

Rita Payan-Carreira Departamento de Matemática da Escola de Ciências e Tecnologia da, Vila Real Universidade de Trás-os-Montes e Alto Douro—UTAD, Vila Real, Portugal

Héctor R. Paz Faculty of Technologies and Exact Sciences, National University of Santiago del Estero, Santiago del Estero, Argentina

Andreas Pester Campus Villach, Carinthia University of Applied Sciences, Spittal, Austria

Clovis António Petry Federal Institute of Santa Catarina, Florianópolis, Brazil

Miguel Plano Faculty of Engineering and Exact Sciences, National University of Rosario, Rosario, Santa Fe, Argentina

María I. Pozzo Research Institute of Rosario in Educational Sciences, National Scientific and Technical Research Council, Santa Fe, Argentina

Bárbara Rangel Mestrado em Design Industrial e de Produto, Design Studio FEUP, Porto, Portugal

Elio SanCristóbal-Ruiz Department of Control, Electrical and Electronics Engineering, National Distance Education University, Madrid, Spain

Luis C. Schlichting Department of Electronics, Federal Institute of Education, Science and Technology of Santa Catarina, Florianópolis, Santa Catarina, Brazil

M. Teresa Sena Esteves Centro de Inovação em Engenharia e Tecnologia Industrial, Instituto Superior de Engenharia do Porto (ISEP), Porto, Portugal

Helena Silva Departamento de Matemática da Escola de Ciências e Tecnologia da, Vila Real Universidade de Trás-os-Montes e Alto Douro—UTAD, Vila Real, Portugal

Juarez B. da Silva Federal University of Santa Catarina, Mato Alto, Araranguá, Santa Catarina, Brazil

Filomena Soares Centro ALGORITMI, School of Engineering, University of Minho, Guimarães, Portugal

Mario F. Soria Faculty of Technologies and Exact Sciences, National University of Santiago del Estero, Santiago del Estero, Argentina

Guilherme Temporão Department of Electrical Engineering, Pontifical Catholic University of Rio de Janeiro, Gávea, Rio de Janeiro, Brazil

Paulo Vasco Departamento de Matemática da Escola de Ciências e Tecnologia da, Vila Real Universidade de Trás-os-Montes e Alto Douro—UTAD, Vila Real, Portugal

Maria C. Viegas School of Engineering, Centre for Innovation in Engineering and Industrial Technology, Polytechnic of Porto, Porto, Portugal

Johan Zackrisson Department of Signal Processing, Blekinge Institute of Technology, Karlskrona, Sweden

Danilo G. Zutin Carinthia University of Applied Sciences, Spittal, Austria

Introduction

Maria M. Nascimento[2], Gustavo R. Alves
and Eva Virgínia Araújo Morais

The Portuguese Society for Engineering Education

As stated in its Web page [1]

"The Portuguese Society for Engineering Education [acronym SPEE] is a non-profit association that aims to promote engineering education through pedagogical training and personal development of teachers, dissemination, and collaboration in projects, exchange between people and national and foreign institutions and analysis and problems in the field of engineering education. Constituted by a public deed signed by a seminal group of founding members, it was launched (...) in 2010. It currently has 20 institutional partners, including the Portuguese Order Of Engineers, almost all university engineering schools and a significant group of polytechnics, and more than 200 individual partners working in an even wider range of schools, in a wide range of fields, from the most classic to the most modern."

Hence, a group of engineers that are teaching in different engineering schools and degrees in Portugal recognized that something more could be done to improve engineering education and SPEE was the way to gather efforts in order to be successful.

The International Conference CISPEE

SPEE devised an international conference to disseminate work and research done and to share views with each other. Thus, the international conferences—acronym CISPEE—arose and are dedicated to engineering education, and its goals are to become a major discussion forum for all stakeholders groups of engineering education. At the same time, CISPEE aims to gather academics, researchers, and

---

[2] Universidade de Trás-os-Montes e Alto Douro—UTAD, University of Trás-os-Montes e Alto Douro, www.utad.pt, Vila Real—Departamento de Matemática, Polo II da ECT, Quinta de Prados, 5000-801,Vila Real, Portugal, Email: mmsn@utad.pt.

professionals directly or indirectly linked to engineering education in order to discuss the progress in this work field and to disseminate its results and each new approach. Usually, CISPEE includes pre-conference workshops, plenary, thematic, and posters sessions. The authors are invited to present their work in the major conference topics: Engineering Ethics; Information and Communications Technologies (ICT) in Engineering Education; Continuing Engineering Education (CEE); Mathematics in Engineering Education; Tools to Develop Higher Order Thinking Skills; and Future Outlook of the Profession and Education in Engineering.

In 2013, this society organized the First International Conference of the Portuguese Society for Engineering Education (CISPEE 2013) held at ISEP, Polytechnic of Porto, Porto, Portugal, and the Second International Conference of SPEE was in 2016 (CISPEE 2016) and was held at University of Trás-os-Montes and Alto Douro (UTAD) in Vila Real, Portugal.

CISPEE 2016 edition brought together teachers and researchers from several engineering schools, from Portugal, and from the international community (e.g., Canada, Spain, South America, and Europe) to share good practices that may contribute to *(Re)Thinking Higher Engineering Education*, the issues related to critical thinking and problem-solving, communication, collaboration, creativity, and innovation in engineering education.

(Re)Thinking Higher Engineering Education was and still is challenging since engineering education is the activity of teaching knowledge and principles—higher education—related to the professional practice of engineering—life and life-long learning. Therefore, beyond the examination of the economic, cultural, and social factors, which influence the education of engineers in different higher education institutions, we should question ourselves about critical thinking and problem-solving, communication, collaboration, creativity, and innovation provided to engineering students (the four C referred in the 2010 American Management Association survey [2]). Training those skills in higher education students may change the way they look at issues, organize their views, and incorporate others' views in order to stimulate new perspectives and prevent biased views of a real situation or problem.

Higher Education in Portugal

Concisely, the Bologna Process (BP) in the context of the European institutions of higher education implied changes in aspects related to the role of the student, the process of learning and training, and the matrix of learning assessment. Underlying BP is the paradigm that emphasizes student centrality, focusing on learning and on its active role [3].

As stated in the Decree-Law no. 74/2006 of March 24, in Portugal, the main objectives were "to guarantee Bologna, a unique opportunity to encourage higher education, to improve the quality and relevance of the courses offered, to promote the mobility of our students and graduates, and to internationalize our courses." In 2008, the "Basic Law of the Educational System" established a model of organization by cycles where the "degrees [1st cycle], master [2nd cycle], and doctorate

[3rd cycle]" are given, as set in the Bologna Declaration. Another important concept is the European Credit Transfer System (ECTS), which translates the "unit of measurement of student work" (Decree-Law no.107/2008 of June 25 on the award of Degrees and Diplomas). The aim is to achieve one of the goals of the BP, that is the transition from the transmission of knowledge to a system centered on training and learning, and so, in addition to having to master the technical skills of a specialized area, it must be able to communicate, lead, work as a team, analyze, and solve problems. This approach requires an experimental component, a problem-based component, and others, and the acquisition of transversal skills should play a major role [3].

The Structure of this Book

In Portugal, the BP was implemented in higher education in 2010. Therefore, a parallel may be set between the engineering educators' concerns and the need they felt to associate—them and their surrounding institutions—in SPEE. As already mentioned, the appearance of CISPEE was almost immediate. At the two CISPEE editions, the committees—Scientific and Organizing—delegate in a jury the choice of the best paper and the best poster; thus, this year we decided to invite the best papers' authors to write a chapter for a book, since the publishing company accepted our proposal. We thought it was worthwhile to gather the best works of the authors under a larger theme of CISPEE 2016 (Re)Thinking Higher Engineering Education, so in the *Contributions to Higher Engineering Education* was not possible to include all the papers presented at CISPEE 2016 on the book.

The first chapter won the price of best paper and presents an "International Cooperation for Remote Laboratory Use," authored by 44 collaborators of the VISIR+ Project. The chapter presents the cooperation needed to implement a remote laboratory, named Virtual Instruments System in Reality (VISIR), in order to provide an additional experimental component in electrical and electronics engineering.

The second chapter studies "Mature Learners' Participation in Higher Education and Flexible Learning Pathways: Lessons Learned from an Exploratory Experimental Research." This chapter discusses a European and Portuguese problem that is the inclusion of a new kind of students in higher education, the mature students, and an alternative way to deal with the gaps those students present.

The third chapter presents "The Flow of Knowledge and Level of Satisfaction in Engineering Courses Based on Students' Perceptions." Based on a survey to students in higher education in two countries, some perception variables are analyzed in light of students' perceptions.

The fourth chapter discusses "Innovative Methodologies to Teach Materials and Manufacturing Processes in Mechanical Engineering." Again, the experimental and the problem-based component as well as other methodologies are implemented in order to engage the students in the classes.

The fifth chapter discusses "'Learning by Doing' Integrated Project Design in a Master Program on Product and Industrial Design." The chapter is about the problem-based implementation in a master in a design study scenario in connection with the industry and real clients.

The sixth and last chapter presents and discusses "The Views of Engineering Students on Creativity." In line with the needs of the twenty-first-century interpersonal, applied skills—creativity included—required also to engineers, an exploratory study is presented about the definitions about creativity of first-year students in two academic years.

In summary, involving the development of experimental components, we have Chapters "International Cooperation for Remote Lab Use" and "Innovative Methodologies to Teach Materials and Manufacturing Processes in Mechanical Engineering," and issues related to topics of the students in the remaining chapters: mature learners' participation and flexible training pathways, in Chapter "Mature Learners' Participation in Higher Education and Flexible Learning Pathways: Lessons Learned from an Exploratory Experimental Research"; flow of knowledge and level of satisfaction in Chapter "The Flow of Knowledge and Level of Satisfaction in Engineering Courses Based on Students' Perceptions"; "learning by doing" in practice product and industrial design in Chapter "'Learning by doing' Integrated Project Design in a Master Program on Product and Industrial Design"; finally, the views of engineering students on creativity in Chapter "The Views of Engineering Students on Creativity."

In order to disseminate information about all the authors devoted to the making of this book, at the end of each chapter we present their short biographic notes.

Conclusion

Contributions to Higher Engineering Education hopes that the reader will be also engaged in a better engineering education. We decided to write about our own experiences and researches in order to leave our testimony to other colleagues and spread the word. It is possible to share our concerns with others, it is possible to cooperate with others, and it is possible to do it in a different way: *The sky is the limit!*

In Isaac Asimov words [4]

"Science can amuse and fascinate us all, but it is engineering that changes the world."

We must continue to involve ourselves as teachers, as engineers, as stakeholders in order *"to promote engineering education through pedagogical training and personal development of teachers, dissemination, and collaboration in projects, exchange between people and national and foreign institutions and analysis and problems in the field of engineering education."*

References

1. Sociedade Portuguesa para a Educação em Engenharia [Portuguese Society for Engineering Education] (2011) http://spee.org.pt/. Accessed 15 June 2017
2. American Management Association (2010) AMA 2010 critical skills survey. Executive summary. Retrieved October www.amanet.org. Accessed 15 June 2017
3. Fernandes SR, Flores MA and Lima RM (2010) A aprendizagem baseada em projectos interdisciplinares: avaliação do impacto de uma experiência no ensino de engenharia. [The learning based on interdisciplinary projects: evaluation of the impact of an experience in engineering teaching]. Avaliação: Revista da Avaliação da Educação Superior [Assess.: J. High. Educ. Assess.], 15(3)
4. Gaither CC, Cavazos-Gaither AE (1998). Practically speaking: a dictionary of quotations on engineering, technology and architecture. CRC Press, p 94 [Isaac Asimov's Book of Science and Nature Quotations (1988), p 78]

International Cooperation for Remote Laboratory Use

Gustavo R. Alves, André V. Fidalgo, Maria A. Marques,
Maria C. Viegas, Manuel C. Felgueiras, Ricardo J. Costa,
Natércia Lima, Manuel Castro, Gabriel Díaz-Orueta,
Elio SanCristóbal-Ruiz, Felix García-Loro, Javier García-Zubía,
Unai Hernández-Jayo, Wlodek J. Kulesza, Ingvar Gustavsson,
Kristian Nilsson, Johan Zackrisson, Andreas Pester, Danilo G. Zutin,
Luis C. Schlichting, Golberi Ferreira, Daniel D. de Bona,
Fernando S. Pacheco, Juarez B. da Silva, João B. Alves,
Simone Biléssimo, Ana M. Pavani, Delberis A. Lima,
Guilherme Temporão, Susana Marchisio, Sonia B. Concari,
Federico Lerro, Gaston S. de Arregui, Claudio Merendino,
Miguel Plano, Rubén A. Fernández, Héctor R. Paz, Mario F. Soria,
Mario J. Gómez, Nival N. de Almeida, Vanderli F. de Oliveira,
María I. Pozzo, Elsa Dobboletta and Brenda Bertramo

Abstract Experimenting is fundamental to the training process of all scientists and engineers. While experiments have been traditionally done inside laboratories, the emergence of Information and Communication Technologies added two

"This project has been funded with support from the European Commission. This publication reflects the views only of the authors, and the Commission cannot be held responsible for any use which may be made of the information contained therein".

G. R. Alves (✉) · A. V. Fidalgo · M. A. Marques · M. C. Viegas
M. C. Felgueiras · R. J. Costa · N. Lima
School of Engineering, Centre for Innovation in Engineering and Industrial Technology,
Polytechnic of Porto, Rua Dr. António Bernardino de Almeida 431, 4249-015 Porto, Portugal
e-mail: visirplus@isep.ipp.pt; gca@isep.ipp.pt

A. V. Fidalgo
e-mail: anf@isep.ipp.pt

M. A. Marques
e-mail: mmr@isep.ipp.pt

M. C. Viegas
e-mail: mcm@isep.ipp.pt

M. C. Felgueiras
e-mail: mcf@isep.ipp.pt

R. J. Costa
e-mail: rjc@isep.ipp.pt

1

M. Castro · G. Díaz-Orueta · E. SanCristóbal-Ruiz · F. García-Loro
Department of Control, Electrical and Electronics Engineering, National Distance Education
University, Calle de Bravo Murillo, 38, 28015 Madrid, Spain
e-mail: mcastro@ieec.uned.es

G. Díaz-Orueta
e-mail: gdiaz@ieec.uned.es

E. SanCristóbal-Ruiz
e-mail: elio@ieec.uned.es

F. García-Loro
e-mail: fgarcialoro@ieec.uned.es

J. García-Zubía · U. Hernández-Jayo
Faculty of Engineering, University of Deusto, Avda. de las Universidades, 24, 48007 Bilbao,
Spain
e-mail: zubia@deusto.es

U. Hernández-Jayo
e-mail: unai.hernandez@deusto.es

W. J. Kulesza · I. Gustavsson · K. Nilsson · J. Zackrisson
Department of Signal Processing, Blekinge Institute of Technology, Campus Grasvik,
371 79 Karlskrona, Sweden
e-mail: wlodek.kulesza@bth.se

I. Gustavsson
e-mail: ingvar.gustavsson@bth.se

K. Nilsson
e-mail: kns@bth.se

J. Zackrisson
e-mail: johanz@gmail.com

A. Pester · D. G. Zutin
Campus Villach, Carinthia University of Applied Sciences, Villacher Straße 1,
A-9800 Spittal, Austria
e-mail: pester@fh-kaernten.at

D. G. Zutin
e-mail: dgzutin@ieee.org

L. C. Schlichting · G. Ferreira · D. D. de Bona · F. S. Pacheco
Department of Electronics, Federal Institute of Education, Science and Technology of Santa
Catarina, Av. Mauro Ramos, 950 – Centro, 88020-300 Florianópolis, Santa Catarina, Brazil
e-mail: schlicht@ifsc.edu.br

G. Ferreira
e-mail: golberi@ifsc.edu.br

D. D. de Bona
e-mail: dezan@ifsc.edu.br

F. S. Pacheco
e-mail: fspacheco@ifsc.edu.br

J. B. da Silva · J. B. Alves · S. Biléssimo
Campus Araranguá, Federal University of Santa Catarina, Rua Pedro João Pereira, 150,
88900-000 Mato Alto, Araranguá, Santa Catarina, Brazil
e-mail: juarez.silva@ufsc.br

J. B. Alves
e-mail: joao.bosco.mota.alves@ufsc.br

S. Biléssimo
e-mail: simone.bilessimo@ufsc.br

A. M. Pavani · D. A. Lima · G. Temporão
Department of Electrical Engineering, Pontifical Catholic University of Rio de Janeiro, Rua
Marquês de São Vicente, 225, 22451-900 Gávea, Rio de Janeiro, Brazil
e-mail: nmm@isep.ipp.pt

A. M. Pavani
e-mail: apavani@puc-rio.br

D. A. Lima
e-mail: delberis@ele.puc-rio.br

G. Temporão
e-mail: temporao@opto.cetuc.puc-rio.br

S. Marchisio · S. B. Concari · F. Lerro · G. S. de Arregui · C. Merendino · M. Plano
Faculty of Engineering and Exact Sciences, National University of Rosario, Maipú 1065,
S2000CGK Rosario, Santa Fe, Argentina
e-mail: smarch@fceia.unr.edu.ar

S. B. Concari
e-mail: sconcari@fceia.unr.edu.ar

F. Lerro
e-mail: flerro2@yahoo.com.ar

G. S. de Arregui
e-mail: gsaez218@gmail.com

C. Merendino
e-mail: claudiomerendino@hotmail.com

M. Plano
e-mail: mplanoster@gmail.com

R. A. Fernández · H. R. Paz · M. F. Soria · M. J. Gómez
Faculty of Technologies and Exact Sciences, National University of Santiago del Estero,
2100, Av. Belgrano Sur, 4200 Santiago del Estero, Argentina
e-mail: raf@unse.edu.ar

H. R. Paz
e-mail: hrpazunse@unse.edu.ar

M. F. Soria
e-mail: fernandosoria@unse.edu.ar

M. J. Gómez
e-mail: mariog76@hotmail.com

N. N. de Almeida · V. F. de Oliveira
Brazilian Association of Engineering Education, SRTVN Quadra 701, Conjunto C, Centro
Empresarial Norte Bloco A, Salas 730/732, 70719-903 Brasília, DF, Brazil
e-mail: nivalnunes@yahoo.com.br

alternatives accessible anytime, anywhere. These two alternatives are known as virtual and remote laboratories and are sometimes indistinguishably referred as online laboratories. Similarly to other instructional technologies, virtual and remote laboratories require some effort from teachers in integrating them into curricula, taking into consideration several factors that affect their adoption (i.e., cost) and their educational effectiveness (i.e., benefit). This chapter analyzes these two dimensions and sustains the case where only through international cooperation it is possible to serve the large number of teachers and students involved in engineering education. It presents an example in the area of electrical and electronics engineering, based on a remote laboratory named Virtual Instruments System in Reality, and it then describes how a number of European and Latin American institutions have been cooperating under the scope of an Erasmus+ project, for spreading its use in Brazil and Argentina.

Keywords Engineering education · Remote laboratories · VISIR
Community of practice · Online laboratories federation

1 Introduction

Remote laboratories stand for physical apparatus connected to computer-controlled instruments which are able to be remotely accessed for carrying out real-world experiments. This definition leads to the expression "remote experimentation" which denotes the type of experiments that can be done in remote laboratories, in opposition to "virtual experiments," or "simulations," which can be done in "virtual laboratories." For a complete understanding, hands-on laboratories refer to physical spaces where users perform experiments by directly manipulating the instruments and/or apparatus under experimentation. The more recent expression "hybrid laboratories" refers to a sort of environment where parts of the apparatus under experimentation and/or the instruments connected to those apparatus are real, and other parts are modeled, i.e., correspond to mathematical and data models running on a computer. These two parts interact during the course of an experiment, hence the word "hybrid."

V. F. de Oliveira
e-mail: vanderli.fava@ufjf.edu.br

M. I. Pozzo · E. Dobboletta · B. Bertramo
Research Institute of Rosario in Educational Sciences, National Scientific and Technical Research Council, Ocampo y Esmeralda, Predio CONICET Rosario, Santa Fe, Argentina
e-mail: pozzo@irice-conicet.gov.ar

E. Dobboletta
e-mail: elsadobboletta@gmail.com

B. Bertramo
e-mail: brendabertramo@gmail.com

In historical terms, the value of experimentation in science has long been recognized. For instance, the oldest Scientific Society in the world, the Royal Society, adopted the motto 'Nullius in verba' to "... *express the determination of its Fellows ... to verify all statements by an appeal to facts determined by experiment*" [1]. This spirit has also long been part of the training process of both scientists and engineers, as reported by Feisel and Rosa (2005) in [2]. In particular, these authors trace back the value of combining theory and practice to the very first engineering school in the USA, the US Military Academy, founded at West Point, NY, in 1802 [2, p. 122]. Although majorly focusing on the role of hands-on laboratories in undergraduate engineering education, Feisel and Rosa (2005) also account for the provisions of both virtual and remote laboratories to that role.

The particular aspects of combining hands-on, simulated, and remote laboratories into Science, Engineering, Technology and Mathematics (STEM) education are well discussed in [3–5]. These papers also acknowledge virtual and remote laboratories to be the two most recent environments where students may acquire and practice some of their experimental competences. Froyd et al. (2012) corroborate this statement by rightfully classifying simulations and remote laboratories as part of one of the five major shifts in 100 years of engineering education, in particular of its fifth major shift, i.e., the influence of Information and Communication Technologies (ICT) in engineering education [6].

But while the generalized use of simulations in engineering education followed the widespread use of computers (1970s), remote laboratories have a more recent history, mainly powered by the emergence of the World Wide Web (WWW) (1990s) [7]. Other aspects impairing the large adoption of remote laboratories, when compared to virtual laboratories, are the associated development, maintenance costs, and scalability constraints [8]. In this chapter, we first briefly expand on this problem and then present one strategy for spreading the use of remote laboratories in Brazil and Argentina, through an international cooperation project. This project involves a number of European and Latin American higher education institutions and is supported by the Erasmus+ program, under the Capacity Building in Higher Education action.

The remainder of the chapter is structured as follows: Sect. 2 provides some background on the use of virtual versus remote laboratories, while also defining one particular application domain—experiments with electrical and electronic circuits; Sect. 3 focuses on one particular remote laboratory serving this domain; Sects. 4 and 5 deal with two crucial aspects for spreading the use of remote laboratories, i.e., nurturing a strong community of practice (CoP) and federating existing remote laboratories; Sect. 6 presents two ongoing international projects around one particular remote laboratory: one project aiming to spread its community of practice in Brazil and Argentina and the other aiming to federate a number of existing nodes in Europe; and, finally, Sect. 7 presents the conclusions and future perspectives.

2 Background

2.1 Scalability Constraints

One possible direction for analyzing the scalability problem of virtual versus remote laboratories is to look into the dimension and hierarchical structure of an engineering school or faculty, while focusing on the practical educational component. At the very basis, one has a single experiment. The dimensional aspect can be reduced to 1:n for simplicity purposes. Regarding hierarchy, one can consider: experiment—laboratory—course—degree—school—institution. Typically, n experiments are done in a laboratory, usually within a specific scientific domain or sub-domain; e.g., an electrical machines laboratory may accommodate basic electromagnetic experiments to demonstrate the basic principles of electrical machines, such as generators/motors (machines with rotating or moving parts) and transformers (non-rotating machine) to more specific experiments such as the electric efficiency of a motor coupled to a generator, or linear induction motors. A laboratory can then support one course or several courses. Those courses can be part of a single degree or belong to different degrees. An engineering school usually offers several degrees, e.g., mechanical engineering, electrical engineering, civil engineering, or chemical engineering, among other engineering degrees. Finally, one institution may have one single engineering school or several ones, depending on its dimension. An example could be a traditional university in Europe, located in a single city, with a single campus, or—in opposition—a federal university, in Brazil, with campuses located in different cities pertaining to the same state. Table 1 summarizes this simple overview.

Another dimensional aspect concerns the size of each heading, e.g., the student population attending one degree. One engineering school may offer more traditional degrees, e.g., electrical engineering with a numerus clausus of 1–2 hundred, alongside with more specific degrees, e.g., mechatronics or engineering cybernetics, which may just admit 20–30 new students every year. An example of this heterogeneous scenario is described in Marques et al. (2014), which analyzes application case studies of a particular remote laboratory [11]. In specific, the topic covered by that remote laboratory lasts from a minimum of 3 to a maximum of 14 weeks, while the number of students enrolled in the different courses ranges from 47 to 574 [11, p. 153].

This brief analysis paves the way to the scalability problem of virtual versus remote laboratories. While, for instance, one of the most widely known virtual laboratories in the whole world, i.e., the PhET Interactive Simulations, from the University of Colorado, US, reports (in 2013) over one hundred million (100,000,000) simulations done, after a period of approximately 10 years [12],[1] a

---

[1]The PhET Interactive Simulations Web site, located at https://phet.colorado.edu, reports 360 million accesses in 06.06.2016.

Table 1 A simple overview of the dimension and hierarchy levels related to engineering schools

	1	*n*
Experiment	May range from a few minutes to a complete class. Usually the number of experiments done in a single class depends on the degree year. Initial years may accommodate more experiments due to their relative simplicity, and more advanced years may imply experiments that take more time to complete. The situation of experiments taking more than one class to complete is rare	A set of experiments may form one class (one laboratory script), span over two or more classes, or form one comprehensive module about a specific topic (e.g., "Introduction to DC circuits" may have 10–15 experiments that will take approximately 2–4 weeks to complete). One module may take more or less time depending of being part of the core scientific degree area or not
Laboratory	A laboratory may serve one course or several courses depending on its level (basic, intermediate, advanced). An example of a basic laboratory could be one allowing introductory experiments with electric circuits. An example of an advanced laboratory could be an "OptoElectronic Lab"	Although sometimes several laboratories are needed to support one single course (large number of classes, classes from different courses requiring the same laboratory, etc.), the usual situation is that a single degree often requires the support of several unique laboratories
Course	The basic "educational unit" in many educational institutions. Each course typically comprises a number of contact hours, divided into theoretical and practical ones, and non-contact hours	In a typical semester scenario, each degree usually comprises 4–6 courses, depending on the number of European Credit Transfer System (ECTS) units
Degree	One degree may range from 6 semesters (180 ECTS) to 4 semesters (120 ECTS) depending on its level: undergraduate (B.Sc.) or graduate (M. Sc.). The number of students attending one degree varies quite much, depending on its scope (general, specific) and its level. Taking the example of the Polytechnic of Porto—School of Engineering, one degree may admit 20 new students (e.g., M.Sc. in Computing Engineering and Medical Instrumentation) or 210 (e.g., B.Sc. on Informatics Engineering)	The number of degrees offered, in simultaneous, by a single school depends upon several factors: geographical location, institutional history, type of institution (e.g., university/polytechnic), etc. Taking the same example, the Polytechnic of Porto —School of Engineering offers 14 degrees (undergraduate) and 12 masters (graduate). It is the number of degrees running at the same time that provides an idea of the school size, i.e., number of students, teachers, staff, laboratories
School	A school's size varies quite significantly. Taking the total graduate engineering enrollment numbers published in [9], it may range from 88 (Baylor University, Waco, TX, ranked #118) to 7,504 (Georgia Institute of Technology, Pasadena, CA, ranked #7), which means a scale factor of 85	One institution may have one or more engineering and/or technological schools. One possible example comes from the Polytechnic University of Catalonia (UPC), Spain, which aggregates 12 STEM-related schools [10]

particular remote-controlled laboratory, considered the best one in its category,[2] registered thirteen thousand accesses (13,000) in 2015, for a period of approximately 8 years [13]. To make it comparable, one user access to the Virtual Instruments System in Reality (VISIR) usually accounts for 1–10 experiments; i.e., every time a user clicks on the "Perform Experiment" button, one real, remote experiment is done; hence the total number of experiments may be around 100,000 for the recorded number of accesses. Additionally, the numbers reported in [13] refer to four different VISIR nodes (i.e., servers), while the PhET Interactive Simulations are delivered through a single Web location. Finally, VISIR supports remote experiments with electrical and electronic circuits (one specific topic, within electrical and electronics engineering), while PhET Interactive Simulations cover several scientific domains like physics, chemistry, biology, and mathematics, among others.

Although the observable simulated-to-remote experiment ratio of this example (in the range of 1:1,000) may be considered as just one possible case, non-representative of all possible comparative cases, the fact is that one simulation corresponds to running a given number of code lines, which can either occur at the server or client side, depending on the technology used. A server with a processing power of hundreds to thousands of Millions of Instructions Per Second (MIPS) can thus deliver many simulations per second, whereas the time duration of a remote experiment is dictated by its physical nature. In the electrical and electronic domain, these experiments may typically take less than a second to complete [14]. However, one may quickly think of other experiments that may take several minutes to complete in the real world (e.g., check relationships between volume and amount of solute to solution concentration) and only a few seconds to simulate (e.g., https://phet.colorado.edu/en/simulation/concentration).

A final note on this topic concerns the access/delivery type of a remote versus a virtual laboratory. A remote laboratory can either work interactively or in a batch mode [15]. In the batch mode, a remote laboratory receives a request from a user, setups the experiment, runs it, and then sends the result(s) to the user. In the interactive mode, one single user is in control of the entire laboratory for the duration of a pre-defined time slot. There are remote laboratories that work on the batch mode, interactive mode, or both. An example would be a remote telescope. In the batch mode, a user defines a particular set of coordinates and filter lens and submits the request to the laboratory. The laboratory will accommodate the request on the first possible time frame and then send the result(s) to the user. In the interactive mode, a user will remotely control the telescope for, e.g., one hour, changing its parameters in real-time, and obtaining the results in real-time.

---

[2]According to the Global Online Laboratory Consortium (GOLC), which granted this award, on its 1st edition (2015), to VISIR.

2.2 Development and Maintenance Costs

The topic of development and maintenance costs, applied to virtual and remote laboratories, may be divided into its software and hardware components. While a virtual laboratory typically consists of a server (the hardware component) running the simulations (the software component), a remote laboratory may include more than one server, as illustrated in [16], the whole experimental apparatus—these two parts forming the hardware component—and the several software layers that form the interface between the remote user and the apparatus under experimentation. The larger number of parts forming the hardware and software components of a remote laboratory and the possible existence of consumables are just two supporting evidences that remote laboratories present higher development and maintenance costs than virtual laboratories.

These higher costs, however, sustain the advantages of using remote experiments, in opposition to just using simulations, in the following cases:

- Simulation results may be quite different from the results of real physical experiments; for instance, in mechanical engineering influences like vibration, torque, and friction cannot be studied and understood so well.
- In order to approach simulation results to real physical experiment results, developers try to improve the accuracy of mathematical and data models. However, this effort has two main drawbacks: (i) It implies higher development costs and (ii) higher computational power, either from the server or the client side. Concerning (i), one could consider the cost of placing online a simple real experiment of a driving motor coupled to a load, versus the cost of developing the most accurate model accounting for all physical variables present in this system and its environment (temperature, humidity, etc.). Respecting (ii), presenting m-learning scenarios, i.e., the use of mobile handheld devices for teaching and learning purposes, still do not account for the possibility to run highly demanding applications, for two main reasons: (a) Very large applications require too much memory and time to download, and (b) handheld devices often present less computational power than portable computers.

On the other side, there are areas where the whole development is based on simulations (e.g., Systems-on-a-Chip), and the real experiments are the way to test, but not to develop. So in any case, both methods are needed in any solid engineering training.

3 The VISIR System

VISIR is a remote laboratory for wiring and performing experiments with electrical and electronic circuits. Basically, it replicates a laboratory workbench equipped with a digital multimeter, a triple DC power supply, a two-channel oscilloscope, a signal generator, and a solderless breadboard, similar to the one illustrated in Fig. 1. This sort of workbench is similar in all engineering schools and faculties, for experimenting electrical and electronic circuits. A stack of boards acting simultaneously as a component store and a reconfigurable matrix, able to interconnect the components and the test and measurement instruments, emulates the solderless breadboard. Figure 2 depicts a VISIR system based on PXI instruments. The basic characteristics of the VISIR platform were initially described by Gustavsson (2001) in [17] and then further explained in [18–24].

If we consider the remote laboratory itself, we can highlight some innovative aspects. Based on the interaction of a simulation of real equipment and real instruments at distance, VISIR creates a real electronic laboratory environment to the student, which can be accessed at anytime and from anywhere as long as the student has a PC connected to Internet [25]. Within such environment, students interact with real instruments and electric/electronic components. They adjust the instruments and wire the circuits with their PC mouse; then, the laboratory sends the measurement results to them, on their PC screen. Students can also control stimulus (e.g., power supply voltages and input signals), using the PC mouse.

As a platform system, VISIR has its own Web interface in which the laboratory contents are arranged and through which they are accessed. It contains many access and administration features such as: registration, login, booking, account types. The availability of the laboratory contents depends on the user account type. Each user account type has its own features, privileges, and limits. Some universities have integrated VISIR into their own Learning Management System (LMS), and/or their

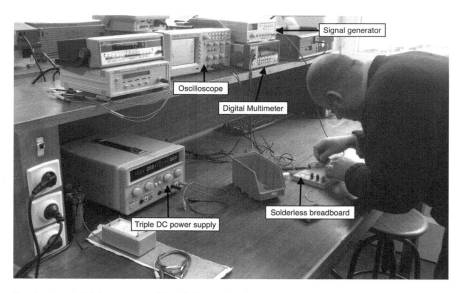

Fig. 1 A typical laboratory workbench for performing experiments with electrical and electronic circuits

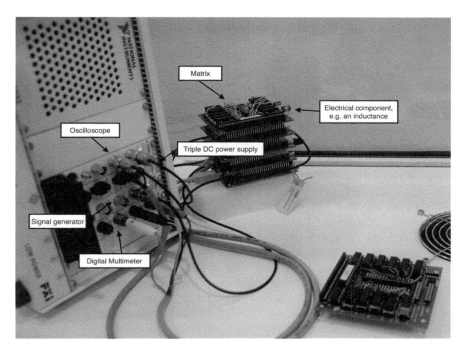

Fig. 2 Hardware component of the VISIR system—version based on PXI instruments

Remote Lab Management System (RLMS), allowing the use of the provided LMS services besides the laboratory work to create a rich integrated online educational platform. So, VISIR may be considered as a remote workbench, equipped with the same instruments that exist in a hands-on laboratory for conducting experiments with electric and electronic circuits. These workbenches are similar to each other, no matter of what part of the world they are being used for supporting laboratory classes with such circuits. This means VISIR has a universal and familiar interface that facilitates its usage. Its limited scope comes as an advantage, because all users immediately know what they are interacting with, either being students, teachers, or project partners.

4   Community of Practice (CoP)

In brief, a CoP is a group of people informally bound together by shared expertise, a set of problems, or interest in a topic or fulfillment of goals [26]. In addition, a CoP focuses on sharing best practices and creating new knowledge to advance a domain of professional practice.

The formation of a CoP around VISIR was inspired by general discussions around the following question: "What is the added value of Remote Experimentation to Education?" This question arose in a former collaborative research project named Remote Experimentation Network—Yielding an Inter-university Peer-to-Peer e-service (RexNet-yippee), which involved several higher education institutions (HEI) from Europe and Latin America (LA) [27]. Although not completely answered, this question was partially addressed by the simple equation presented in Fig. 3. In face of the difficulty in reaching a precise quantitative formula able to compute such added value, the proposed qualitative formula was simple enough to point directions on how to increase it. In simple terms, if one increases the educational value of a given remote experiment and, simultaneously, decreases its development and maintenance costs, then the resulting added value will increase.

The two guidelines suggested in the formula for increasing the educational value are in line with the objectives of a CoP. These same guidelines form part of a project proposal (VISIR+) submitted to the Erasmus+ program, for enlarging a CoP around VISIR with European HEI that already have this system and a number of Latin American HEI, which have a rich experience on the use of remote experimentation, but do not have VISIR. An important aspect that needs to be highlighted at this stage is that such collaboration implies a shared knowledge and interest in a given scientific area. In the case of VISIR+, this concerns the teaching and learning of electric/electronic circuits' theory and practice.

The CoP around VISIR actually started as a Special Interest Group (SIG) of the International Association for Online Engineering (IAOE), circa 2006. While initially gathering researchers interested in enhancing and spreading VISIR system [18], it soon started to benefit from the input of a larger number of users, i.e., from teachers and students, and effectively growing into a CoP. The following list presents some of the results achieved by this CoP, in the past 10 years:

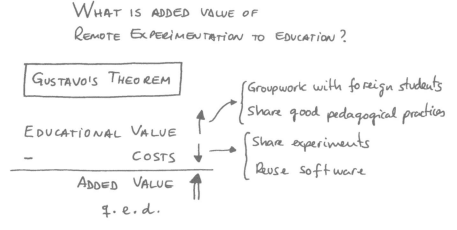

Fig. 3 A simple formula for evaluating the added value of remote experimentation to education [15]

- A number of technical improvements in the VISIR hardware (the relay matrix) and software (the user interface) directly resulting from the received user feedback [11, 28].
- Developing a cheaper and equally reliable platform based on LXI-compatible instruments [29, 30].
- Outreaching a larger number of students and teachers, effectively helping in expanding the existing CoP. So far, approximately 50 teachers and 5,000 students have used VISIR, in particular considering its use in a Massively Open Online Course (MOOC) developed by the Spanish National Distance Education University (UNED) [31].
- Evidence of collaborative episodes involving teachers and students from different world regions, namely from Europe, Latin America, Middle East, and Australia [32–36].

But, in order to effectively support an even larger community, the simple existence of several VISIR nodes is not enough. The reasoning is simple and implies two directions: number of available experiments and number of students and teachers served in simultaneous. Considering all the experiments done with electrical and electronic circuits, in a single semester, it is clear that, even with a large relay matrix, one single VISIR system is unable to serve one single school. Considering a simple, yet widely performed experiment like an RC lowpass filter, it is obvious that one single VISIR system is unable to serve all engineering schools. The solution to this scalability problem is presented in the next section.

5 A Federation of VISIR Nodes

The two other guidelines suggested in the formula for decreasing the development and maintenance costs of remote laboratories are better understood within the conceptual definition of a federation. When sharing experiments, institutions may choose to: (i) simply open their access to anyone hitting the Web page where they are located; (ii) disseminate their existence (and access to) through a repository; or (iii) join a federation that allows some sort of single sign-on (SSO) facility. Examples of (i) are the Control System Online Laboratory, developed by Jim Henry and hosted by the University of Tennessee at Chattanooga, US [37], or any VISIR system, when accessing the demo page and using the guest login [38]. Examples of (ii) are the European Go-Lab portal, which provides access to hundreds of online laboratories [39], or the Lab2Go portal [40]. Finally, examples of (iii) are the Labshare institute [41], or the iLab Service Broker [42]. Unfortunately, option (i) does not really provide any sort of rewarding mechanism, as there is no structured way to access other remote experiments. Although the possibility to search the Web for any particular, open, remote experiment still exists, it is a random, time-consuming process, where the guarantee of a quality of service (e.g., the remote experiment remains open for an entire course duration) is virtually zero.

Option (ii) is more structured and facilitates the task of searching and using a given remote experiment. However, it is up to the owner of the repository to set up the rules defining how a given remote experiment is made publicly available and what sort of service level must be provided. Usually, by joining such a repository, a given institution will have to provide but also be allowed to use remote experiments provided by other institutions. In some cases, the repository is completely open; i.e., all the remote experiments listed in the repository are open, often with some sort of restriction (limited access time, diminished complexity, etc.). Again, this sort of sharing presents more advantages to users rather than to providers; i.e., the two directions (provider–client and client–provider) are not balanced in terms of benefits.

A federation implies a different quality of service level, in relation to a repository. It offers a server or now often cloud-based user and laboratory management in one system. Administrators can define laboratory and user groups, and their roles, and offer pre-defined access types to the online laboratories and remote experiments. Via special Web services (smart gateways [43]), these systems can be connected to an LMS by single sign-on, if the LMS supports the Learning Tools Interoperability (LTI) protocol. The laboratory owner in every case defines the use policy (time frames, actual number of users, etc.) of his laboratory. But he accepts that all (usually identified) users, who are registered into the laboratory group of the federation to which his laboratory is connected to, have access to his experiments.

Orduña et al. (2015) expose the advantages of a federated system [43] concerning the experiments shareability: "*once students of a particular institution can access through the Internet to a particular laboratory, it can also be accessed by students of other universities*". This advantage is bidirectional through RLMSs in which a federation is established: Two institutions providing the same remote laboratory—or the same practical experiment from a specific remote laboratory—can balance their clients/users load. This feature, inherit to RLMSs, improves the users' immersion in the remote laboratory environment due to the improved time response.

Laboratory time response depends on several factors: circuit, frequency, number of measuring requests, etc. In any case, there is a physical constraint to the number of concurrent users performing measurements; threshold limit value is 60 in VISIR [44]. Even though it is unlikely that all connected users perform measurements simultaneously—laboratory time is mostly allocated to circuit assembling and configuring the equipment—much more than for measuring, a balanced users' load for some particular experiments in strong demand would provide a better time response, and hence a better immersion.

This particular aspect is visible through the following sequence of experiments, done with a single VISIR node (Figs. 4, 5, and 6).

The sequence shows the increasing delay in serving an increasing number of simultaneous users, based on the batch operation mode of VISIR. The number of potential users, in a single engineering school, presented in Sect. 2, helps to understand the limitations of having a VISIR node operating in an isolated fashion.

Another approach to build a VISIR federation is to carry it out following a strategical design of the practical experiments offered by the different VISIR nodes.

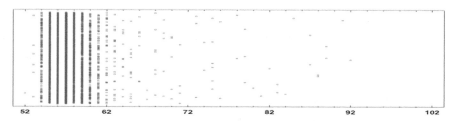

Fig. 4 Unique user, time response in milliseconds; 5 min in continuous mode

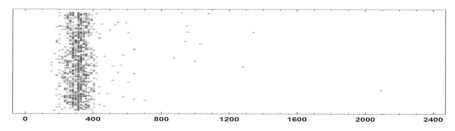

Fig. 5 Five users simultaneously measuring sample time response in milliseconds; 5 min in continuous mode

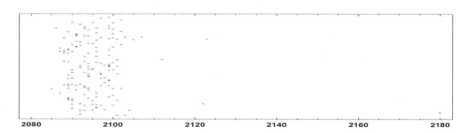

Fig. 6 Over 20 users simultaneously measuring sample time response in milliseconds; 5 min in continuous mode

Every VISIR node of the community could share a "percentage" of its matrix to the VISIR federation. If every VISIR node offers a rich and broad specialized block of experimental practices (i.e., Node 1: basic circuits and electrical laws; Node 2: diodes experimentation; Node 3: transistors experimentation; Node 4: OpAmp experimentation; etc.), the overall VISIR nodes would share a huge and plentiful electronics practices repository, enriching exponentially the availability and quality of practical experiments. This repository could also be extended to practical guides and additional documentation, forming a VISIR community not only for sharing resources but also for a continuous improvement at all levels.

Finally, this whole notion of building a federation of individual nodes is not unique to remote laboratories; rather there are also examples of proposals emerging from the area of simulations, as presented in [45].

6 Ongoing Projects Around VISIR

6.1 VISIR+

The motivation to submit a project proposal for installing new VISIR nodes in Brazil and Argentina emerged from: (1) the history of collaboration around VISIR among a number of European and LA HEI; (2) the current demand for an increased use of instructional technologies in Science and Engineering Education, able to supply these two countries with a better skilled workforce; and (3) the opportunity presented by the Erasmus+ program, favoring joint projects between these two world regions, under the scope of the Capacity Building in Higher Education (CBHE) measure. Under this scope, the Polytechnic of Porto (IPP), from Portugal, the National Distance Education University (UNED) and the University of Deusto (UD), both from Spain, the Carinthia University of Applied Sciences (CUAS), from Austria, the Blekinge Institute of Technology (BTH), from Sweden, the Pontifical Catholic University of Rio de Janeiro (PUC-Rio), the Federal University of Santa Catarina (UFSC), the Federal Institute of Santa Catarina (IFSC), the Brazilian Association for Engineering Education (ABENGE), all the previous 4 institutions from Brazil, the National University of Rosario (UNR), the National University of Santiago del Estero (UNSE), and the Research Institute of Rosario for Educational Sciences (IRICE-CONICET), all the previous 3 institutions from Argentina, joined forces together and submitted a project proposal to the very first call of Erasmus + program, on 10 February 2015. The project proposal, entitled "Educational Modules for Electric and Electronic Circuits Theory and Practice following an Enquiry-based Teaching and Learning Methodology supported by VISIR", and shortly referred as VISIR+, was positively evaluated in July 2015 and had its Kick-Off-Meeting (KOM) in Karlskrona, Sweden, on 1–3 February 2016.

In brief, VISIR+ is installing five new VISIR nodes in the Brazilian and Argentinean HEI, i.e., PUC-Rio, UFSC, IFSC, UNR, and UNSE, with the assistance of the European HEI who already have one or more VISIR systems installed, i.e., BTH, IPP, UNED, UD, and CUAS. IRICE-CONICET is responsible for quality monitoring the didactical implementation of the new VISIR nodes, and ABENGE will support the dissemination and impact evaluation of the VISIR+ project. Figure 7 provides an idea of the geographical distribution of the VISIR+ consortium.

In order to effectively grow the CoP around VISIR, the project includes the following three training actions (TA):

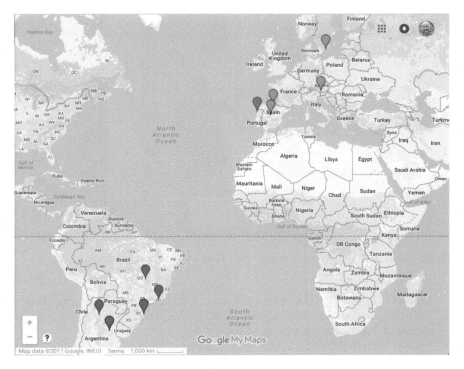

Fig. 7 Geographical distribution of the VISIR+ consortium

- A first one held at BTH, with presentations done by all European partners. Two representatives from the Argentinean and Brazilian HEI participated locally, while an additional number of teachers participated remotely. At the end of this activity, participants were expected to know what VISIR is, what experiments it can support, how it can be incorporated into a course curricula, what learning outcomes does it enables, etc. A snapshot of TA1, delivered at BTH, is shown in Fig. 8.

- A second TA delivered at each LA HEI. Although the initial plan was to deliver this TA after the local installation of a VISIR node, bureaucratic problems impairing the timely acquisition of the necessary equipment, by these institutions, led to the situation where only PUC-Rio used its newly installed VISIR node to support the local TA. However, this constraint did not prevent the delivery of the TA because of the remote nature of VISIR. Instead of using a local system, the trainers remotely used the system installed at their home institution in Europe. The target audience of TA2 was the two local representatives who attended TA1 plus all interested teachers from the same institution and also, at least, one representative from the associated partners. These associated partners—two per LA HEI—are nearby educational institutions also

Fig. 8 Snapshot of TA1 delivered at BTH (1–2 February 2016)

interested in using VISIR. Figures 9, 10, 11, 12, and 13 provide snapshots of TA2 delivered at PUC-Rio, UFSC, UNR, UNSE, and IFSC, respectively. The results of TA2 were reported in [46–48].

- Finally, a third TA to be held at each associated partner. This last TA, jointly delivered by one LA partner and one European partner, will test the capacity to aggregate other institutions around the use of VISIR. This TA will include application examples from courses delivered at the LA HEI, to prove the adaptability of VISIR to different institutional cultures and its universality in terms of experiments with electric and electronic circuits.

Fig. 9 Snapshot of TA2 delivered at PUC-Rio (5–6 September 2016)

Fig. 10 Snapshot of TA2 delivered at UFSC (22–23 August 2016)

Fig. 11 Snapshot of TA2 delivered at UNR (12–16 September 2016)

An underlying common aspect to all TA is the proposed instructional design of all target courses. In particular, VISIR+ aims to develop a set of educational modules comprising the use of hands-on, simulated, and remote laboratories, following an enquiry-based methodology explained in [49–51]. The combination of these three different laboratory environments provides additional opportunities for students to acquire higher-order experimental skills and hence be better prepared to face the labor market [52]. In addition, teachers may use two supplementary tools (simulations and remote laboratories) for enriching theoretical classes, in particular

Fig. 12 Snapshot of TA2 delivered at UNSE (12–16 September 2016)

Fig. 13 Snapshot of TA2 delivered at IFSC (25–26 August 2016)

for proving or demonstrating a given model or formula, which is thought to favor students' motivation and, hence, increase their knowledge retention level [53, 54].

6.2   PILAR

Regarding PILAR, an acronym that stands for "Platform Integration of Laboratories based on the Architecture of VISIR", there are still few results. The project proposal was submitted in February 2016 and positively evaluated in July 2016. The project KOM was held in Madrid, Spain, on November 2016, and the initial activities are now being implemented, in particular a thorough analysis of the characteristics associated with the VISIR systems installed in the consortium partners, i.e., BTH, CUAS, UNED, UD, and IPP. In addition to these partners, the project consortium also includes the International Association for Online Engineering (IAOE), a Small and Medium Enterprise (SME) named EVM Project Management Experts SL, and Omnia, the Joint Authority of Education in Spoo Region, Finland. At the end of the project, it is expected that the first federation of VISIR nodes will be effective and able to provide the services mentioned in Sect. 5.

7 Conclusion and Future Perspectives

Although the two previous projects are still ongoing, some aspects that arise from analyzing the constant growing of the VISIR community should be remarked. The possibility that emerges from a federation of remote laboratories allows sharing resources and widens opportunities for remote experimentation. This means that whereas at a first moment each partner has its own VISIR system, to be used by teachers and students, and shared with other institutions, the next step will be to federate the VISIR systems of the various institutions. What could be achieved from this federation can be described with an example. If the VISIR system of one participating engineering school, located in Argentina, and the VISIR system of another participating engineering school, located in Spain, are integrated into a federation, the students and teachers of those two institutions will have a seamless access to both systems. This is much more than what each institution has developed individually and is able to offer to its teachers and students, alone.

In this way, VISIR+ can be considered the first necessary step to have a federation of VISIR nodes, in which each partner is a supplier and a user at the same time. On its turn, PILAR is the vehicle to implement the first federation of VISIR nodes, in Europe.

Acknowledgements The authors would like to acknowledge the support given by the European Commission through grant 561735-EPP-1-2015-1-PT-EPPKA2-CBHE-JP and also by the Spanish Office for Education Internationalization (Servicio Español para la Internacionalización de la Educación, SEPIE), through grant 2016-1-ES01-KA203-025327, both under the scope of the Erasmus+ program.

References

1. The Royal Society (2016) History. https://royalsociety.org/about-us/history/. Accessed 2 June 2016
2. Feisel LD, Rosa AJ (2005) The role of the laboratory in undergraduate engineering education. J Eng Educ 94:121–130. https://doi.org/10.1002/j.2168-9830.2005.tb00833.x

3. Ma J, Nickerson JV (2006) Hands-on, simulated, and remote laboratories: a comparative literature review. ACM Comput Surv 38(3):1–24. https://doi.org/10.1145/1132960.1132961
4. Corter JE, Nickerson JV, Esche SK, Chassapis C, Im S, Ma J (2007) Constructing reality: a study of remote, hands-on and simulated laboratories. ACM Trans Comput-Hum Interact 14 (2):1–27. https://doi.org/10.1145/1275511.1275513
5. Brinson JR (2015) Learning outcome achievement in non-traditional (virtual and remote) versus traditional (hands-on) laboratories. Comput Educ 87(C):218–237. https://doi.org/10. 1016/j.compedu.2015.07.003
6. Froyd JE, Wankat PC, Smith KA (2012) Five major shifts in 100 years of engineering education. Proc IEEE 100(Special Centennial Issue):1344–1360. https://doi.org/10.1109/ jproc.2012.2190167
7. Alves GR, Gericota MG, Silva JB, Alves JB (2007) Large and small scale networks of remote labs: a survey. In: Gomes L, García-Zubía J (eds) Advances on remote laboratories and e-learning experiences, 1st edn. Universidad de Deusto Press, Bilbao, pp 15–34
8. Orduña P, Rodriguez-Gil L, García-Zubía J, Dziabenko O, Angulo I, Hernández-Jayo U, Azcuenaga E (2016) Classifying online laboratories: reality, simulation, user perception and potential overlaps. In: Proceedings of 13th international conference on remote engineering and virtual instrumentation, Madrid, Spain, 24–26 Feb 2016
9. US News & World Report (2016) Education ranks & advice, US best engineering schools. http://grad-schools.usnews.rankingsandreviews.com/best-graduate-schools/top-engineering-schools/eng-rankings. Accessed 1 July 2016
10. Universitat Politècnica de Catalunya (2016) Own schools. http://www.upc.edu/the-upc/the-institution/schools. Accessed 1 July 2016
11. Marques MA, Viegas MC, Costa-Lobo MC, Fidalgo AV, Alves GR, Rocha JS, Gustavsson I (2014) How remote labs impact on course outcomes: various practices using VISIR. IEEE Trans Educ 57(3):151–159. https://doi.org/10.1109/TE.2013.2284156
12. Khatri R, Henderson C, Cole R, Froyd JE (2013) Over one hundred million simulations delivered: a case study of the PhET interactive simulations. In: Proceedings of the 2013 physics education research conference, Portland, OR, US, 17–18 July 2013
13. Salah R, Alves GR, Abdulazeez D, Guerreiro P, Gustavsson I (2015) Why VISIR? Proliferative activities and collaborative work of VISIR system. In: Proceedings of the 7th international conference on education and new learning technologies, Barcelona, Spain, 6–8 July 2015
14. Swartling M, Bartůněk JS, Nilsson K, Gustavsson I, Fiedler M (2012) Simulations of the VISIR open lab platform. In: Proceedings of the 9th remote engineering and virtual instrumentation conference, Bilbao, Spain, 4–6 July 2012
15. Gomes L, García-Zubía J (eds) (2007) Advances on remote laboratories and e-learning experiences. Universidad de Deusto Press, Bilbao
16. Schmid C (2000) Remote experimentation in control engineering. Paper presented at the 45th international scientific colloquium, Ilmenau Technical University, Germany, 4–6 Oct 2000
17. Gustavsson I (2001) Laboratory experiments in distance learning. Paper presented at the international conference on engineering education, Oslo, Norway, 6–10 Aug 2001
18. Gustavsson I (2003) User defined electrical experiments in a remote laboratory. Paper presented at 2003 annual conference of the American Society for Engineering Education, Nashville, Tennessee, 22–25 June 2003
19. Gustavsson I, Zackrisson J, Håkansson L, Claesson I, Lagö T (2007) The VISIR project—an open source software initiative for distributed online laboratories. In: Proceedings of the 4th remote engineering and virtual instrumentation conference, Porto, Portugal, 25–27 June 2007

20. Zackrisson J, Gustavsson I, Håkansson L (2007) An overview of the VISIR open source software distribution 2007. In: Proceedings of the 4th remote engineering and virtual instrumentation conference, Porto, Portugal, 25–27 June 2007
21. Gustavsson I, Zackrisson J, Ström-Bartunek J, Nilsson K, Håkansson L, Claesson I, Lagö T (2008) Telemanipulator for remote wiring of electrical circuits. In: Proceedings of the 5th remote engineering and virtual instrumentation conference, Düsseldorf, Germany, 22–25 June 2008
22. Gustavsson I, Zackrisson J, Nilsson K, García-Zubía J, Håkansson L, Claesson I, Lagö T (2008) A flexible instructional electronics laboratory with local and remote lab workbenches in a grid. Int J Online Eng 4(2):12–16
23. Gustavsson I (2011) On remote electronics experiments. In: García-Zubía J, Alves GR (eds) Using remote labs in education—two little ducks in remote experimentation. University of Deusto Press, Bilbao, pp 157–176
24. Salah RM, Alves GR, Guerreiro P, Gustavsson I (2016) Using UML models to describe the VISIR system. Int J Online Eng 12(6):34–42. https://doi.org/10.3991/ijoe.v12i06.5707
25. Tawfik M, Sancristobal-Ruiz E, Martin S, Gil M, Pesquera A, Losada P, Diaz-Orueta G, Peire J, Castro M, García-Zubía J, Hernández-Jayo U, Orduña P, Angulo I, Costa-Lobo MC, Marques MA, Viegas MC, Alves GR (2012) VISIR: experiences and challenges. Int J Online Eng 8(1):25–32. https://doi.org/10.3991/ijoe.v8i1.1879
26. Couros A (2003) Communities of practice: a literature review. Available via http://www.tcd.ie/CAPSL/_academic_practice/pdfdocs/Couros_2003.pdf. Accessed 23 Jan 2017
27. Alves GR, Ferreira JM, Müller D, Erbe H-H, Alves JB, Pereira CE, Sucar E, Herrera O, Chiang L, Hine N (2005) Remote experimentation network—yielding an inter-university peer-to-peer e-service. In: Proceedings of the 10th IEEE international conference on emerging technologies and factory automation, Catania, Italy, 19–22 Sept 2005
28. Fidalgo A, Costa-Lobo MC, Marques MA, Viegas MC, Alves GR, García-Zubía J, Hernández-Jayo U, Gustavsson I (2012) Using remote labs to serve different teacher's needs—a case study with VISIR and RemotElectLab. In: Proceedings of the 9th remote engineering and virtual instrumentation conference, Bilbao, Spain, 4–7 July 2012
29. Hernández-Jayo U, García-Zubía J, Angulo I, Lopez-de-Ipiña D, Orduña P, Irurzun J, Dziabenko O (2010) LXI technologies for remote labs: an extension of the VISIR project. Int J Online Eng 6(SI 1):25–35. https://doi.org/10.3991/ijoe.v6s1.1385
30. Hernández-Jayo U, García-Zubía J (2011) A remote and reconfigurable analog electronics laboratory based on IVI an LXI technologies. In: Proceedings of the 8th remote engineering and virtual instrumentation conference, Brasov, Romania, 29 June–2 July 2011
31. García F, Díaz-Orueta G, Tawfik M, Martín S, Sancristobal-Ruiz E, Castro M (2014) A practice-based MOOC for learning electronics. In: Proceedings of the IEEE global engineering education conference, Istanbul, Turkey, 3–5 Apr 2014
32. Ferreira G, Lacerda J, Schlichting L, Alves GR (2014) Enriched scenarios for teaching and learning electronics. In: Atas del XI Congreso de Tecnologías, Aprendizage y Enseñanza de la Electrónica, Bilbao, Spain, 11–13 June 2014
33. Schlichting L, Anderson JA, Faveri F, Bona D, Ferreira G, Alves GR, Liz MB (2016) Remote laboratory: application and usability. In: Atas del XII Congreso de Tecnologías, Aprendizage y Enseñanza de la Electrónica, Seville, Spain, 22–24 June 2016
34. Odeh S, Anabtawi M, Alves GR, Gustavsson I, Arafeh L (2015) Assessing the remote engineering lab VISIR at Al-Quds University in Palestine. Int J Online Eng 11(1):35–38
35. Odeh S, Alves J, Alves GR, Gustavsson I, Anabtawi M, Arafeh L, Jazi M, Arekat M (2015) A two-stage assessment of the remote engineering lab VISIR at Al-Quds University in Palestine. IEEE Revista Iberoamericana de Tecnologias del Aprendizaje 10(3):175–185. https://doi.org/10.1109/RITA.2015.2452752

36. Nafalski A, Göl Ö, Nedić Z, Machotka J, Ferreira JM, Gustavsson I (2010) Experiences with remote laboratories. In: Proceedings of the 7th remote engineering and virtual instrumentation conference, Stockholm, Sweden, 29 June–2 July 2010
37. Laboratories without Borders (2017) Control systems. http://weblab.utc.edu/Weblab/Stations/controlslab.html. Accessed 24 Jan 2017
38. OpenLabs Electronics Laboratory (2017) Guest login. https://openlabs.bth.se/electronics/index.php/en?sel=guestlogin. Accessed 24 Jan 2017
39. The Go-Lab Project (2016) Global online science labs for inquiry learning at school. http://go-lab-project.eu. Accessed 1 July 2016
40. Zutin DG, Auer ME, Maier C, Niederstätter M (2010) Lab2go—a repository to locate educational online laboratories. In: Proceedings of the 1st IEEE global engineering education conference, Madrid, Spain, 14–16 Apr 2010
41. The Labshare Institute (2016) About. http://www.labshare.edu.au/about/institute. Accessed 1 July 2016
42. iLab Service Broker (2016) Welcome to iLab. http://ilab.mit.edu/iLabServiceBroker/. Accessed 1 July 2016
43. Orduña P, Zutin DG, Govaerts S, Lequerica-Zorrozua I, Bailey P, Sancristobal-Ruiz E, Salzmann C, Rodriguez-Gil L, DeLong K, Gillet D, Castro M, López-de-Ipiña D, García-Zubía J (2015) An extensible architecture for the integration of remote and virtual laboratories in public learning tools. Revista Iberoamericana de Tecnologias del Aprendizaje 10(4):223–233. https://doi.org/10.1109/RITA.2015.2486338
44. Orduña P (2013) Transitive and scalable federation model for remote laboratories. Ph.D. dissertation, University of Deusto
45. U.S. Department of Defense (DoD) Modeling and Simulation Coordination Office (M&SCO) (2016) White paper. M&S VV&A RPG Core Document: VV&A of Federations. http://www.msco.mil/documents/Core_Federation01_Federation.pdf. Accessed 11 July 2016
46. Alves GR, Fidalgo A, Marques MA, Viegas MC, Felgueiras MC, Costa RJ, Lima N, Pozzo I, Dobboletta E, García-Zubía J, Hernández-Jayo U, Castro M, García-Loro F, Zutin DG, Christian K (2017) VISIR + project—preliminary results of the training actions. In: Proceedings of the 14th remote engineering and virtual instrumentation conference, New York, NY, USA, 15–17 Mar 2011 (in press)
47. Hernández-Jayo U, García-Zubía J, Colombo AF, Marchisio S, Concari SB, Lerro F, Pozzo MI, Dobboletta E, Alves GR (2017) Spreading the VISIR remote lab along Argentina —the experience in Patagonia. In: Proceedings of the 14th remote engineering and virtual instrumentation conference, New York, NY, USA, 15–17 Mar 2011 (in press)
48. García-Loro F, Fernandez R, Gomez M, Paz H, Soria F, Pozzo MI, Dobboletta E, Fidalgo A, Alves GR, Sancristobal-Ruiz E, Diaz-Orueta G, Castro M (2017) Educational scenarios using remote laboratory VISIR for electrical/electronic experimentation. In: Proceedings of the 14th remote engineering and virtual instrumentation conference, New York, NY, USA, 15–17 Mar 2011 (in press)
49. Lima N, Alves GR, Viegas MC, Gustavsson I (2015) Combined efforts to develop students experimental competences. In: Proceedings of the exp.at'15 conference, Ponta Delgada, Azores, Portugal, 2–4 June 2015
50. Viegas MC, Lima N, Alves GR, Gustavsson I (2014) Improving students experimental competences using simultaneous methods in class and in assessments. In: Proceedings of the technological ecosystems for enhancing multiculturality conference, Salamanca, Spain, 1–3 Oct 2014
51. Alves GR, Viegas MC, Lima N, Gustavsson I (2016) Simultaneous usage of methods for the development of experimental competences. Int J Hum Capital Inf Technol Professionals 7 (1):54–73. https://doi.org/10.4018/IJHCITP.2016010104

52. Restivo T, Alves GR (2013) Acquisition of higher-order experimental skills through remote and virtual laboratories. In: Dziabenko O, García-Zubía (eds) IT innovative practices in secondary schools: remote experiments. Universidad de Deusto, Bilbao, pp 321–347
53. Cooper M, Ferreira JM (2009) Remote laboratories extending access to science and engineering curricular. IEEE Trans Learn Technol 2(4):342–353. https://doi.org/10.1109/TLT.2009.43
54. García-Zubía J, Cuadros J, Romero S, Hernández-Jayo U, Orduña P, Guenaga M, Gonzalez-Sabate L, Gustavsson I (2016) Empirical analysis of the use of the VISIR remote lab in teaching analog electronics. IEEE Trans Educ PP(99):1–8. https://doi.org/10.1109/te.2016.2608790 (in press)

Gustavo R. Alves graduated in 1991 and obtained an M.Sc. and a Ph.D. in Computers and Electrical Engineering in 1995 and 1999, respectively, from the University of Porto, Portugal. He is with the Polytechnic of Porto—School of Engineering since 1994. He was involved in several national and international R&D projects and has authored or co-authored more than 220 publications, including book chapters, and conference and journal papers with a referee process. His research interests include engineering education, remote laboratories, and design for debug and test.

André V. Fidalgo is a researcher at CIETI/LABORIS and has been a lecturer at the Polytechnic of Porto—School of Engineering (IPP/ISEP) since 1999, in Porto, Portugal, where he teaches Electronics and Test and Debug of Digital Systems. His main research interests are related to test and debug of digital systems, remote laboratories, as well the e-learning wider research field.

Maria A. Marques graduated in Physics from University of Porto, received her M.Sc. degree in Physics of Laser Communications from University of Essex in 1992, and her Ph.D. in Engineering Sciences from University of Porto in 2008. She is a Professor at the Polytechnic of Porto—School of Engineering since 1995, where she lectures physics, electronics, and biomechanics courses. She has been involved in several R&D projects and has been a Member of the Physics Department Management Board, in several occasions, since 1999 till present. She is a Member of the IEEE Instrumentation & Measurement Society.

Maria C. Viegas is a Researcher at CIETI/LABORIS and a Professor at the Polytechnic of Porto—School of Engineering, Portugal, since 1994. She graduated in Physics/Applied Mathematics, M.Sc. in Mechanical Engineering, and Ph.D. in Science and Technology (Physics Didactics). Her research interests include engineering education, physics didactics, professional development, and remote experimentation learning outcomes.

Manuel C. Felgueiras received the B.Sc. and Ph.D. degrees in Electrical and Computer Engineering from the Faculty of Engineering, University of Porto, Porto, Portugal, in 1987 and 2008, respectively. He started his activity in 1994 as an Assistant Professor and later on as an Adjunct Professor and Researcher with the Department of Electrical Engineering, School of Engineering, Polytechnic of Porto, Portugal. His research interests at CIETI include design for debug and test of mixed signals, remote experimentation in e-learning, and renewable energy source.

Ricardo J. Costa has a bachelor's degree (1999) and an M.Sc. (2003) in Electrical and Computer Engineering from the Faculty of Engineering of the University of Porto and a Ph.D. (2014) from the University of Coimbra. With more than 25 publications, he has been researching in the area of remote experimentation, and he is now a teacher at the Polytechnic of Porto at the Electrical Engineering Department, teaching electrical and informatics courses, coordinating B.Sc. and M. Sc. degree projects and participating in some international projects.

Natércia Lima is a professor at the Polytechnic of Porto—School of Engineering (ISEP) since 1993, where she teaches Physics, and a full Member of the Research Group in Systems Testing, part of the Center for Innovation in Engineering and Industrial Technology since 2014. Her research interests include engineering education, students learning with remote experimentation, and professional development.

Manuel Castro is an Electrical and Computer Engineering educator in the Spanish University for Distance Education (UNED) and has a doctoral industrial engineering degree from ETSII/UPM. He has been Director of the Department and Vice-Rector at UNED. He has co-chaired the conference FIE 2014 (Frontiers in Education Conference) in Madrid, Spain, by IEEE and the ASEE, as well as REV 2016. He is Fellow Member of IEEE and Past President Sr of the IEEE Education Society. He has been awarded at IEEE EDUCON 2011 and at the IEEE Education Society. He is Co-Editor of IEEE-RITA of the IEEE Education Society and Member of the Board of the Spanish International Solar Energy Society (ISES).

Gabriel Díaz-Orueta has a degree and a Ph.D. in Physics from Universidad Autonoma de Madrid. He is Associate Professor in the Electrical and Computing Engineer Department at Spanish University for Distance Education (UNED). He teaches electronics and information security for networks, and his research is centered in remote laboratories, technology-enhanced learning, and information security. He is the Director of Master Universitario de Investigación en Ingeniería Eléctrica Electrónica y Control industrial at UNED. He is IEEE senior member and past President of the Spanish Chapter of the IEEE Education Society.

Elio SanCristóbal-Ruiz is an Assistant Professor in the Electronic and Computer Department of the Spanish University for Distance Education (UNED). He has a doctoral engineering degree from the Industrial Engineering School (ETSII) of UNED, 2010. He also has a Computer Science Engineering degree from the Salamanca Pontifical University (UPS), Madrid, 2002, and a Technical Engineering degree in Computer Networks (UPS), Madrid, 1998. He has worked for the University Distance Education Institute (IUED) from UNED and in Computer Science Service Centre of UNED.

Felix García-Loro is a Researcher at Spanish University for Distance Education (UNED). He has an Industrial Engineering degree from the UNED and is now pursuing his Ph.D. in Industrial Engineering. He is Assistant Teacher in Electronics for Information and Communication Technologies and Microelectronics and Lecturer in Fundamentals of Electrical Engineering, Medium and Low Voltage Electrical Installations and Electric Machinery. His research interests are focused on integration of remote laboratories, control systems, and big data.

Javier García-Zubía has a degree and a Ph.D. in Informatics. He is a Full Professor in the Faculty of Engineering of the University of Deusto, Spain, where he teaches electronics and embedded systems and investigates in remote laboratories, technology-enhanced learning, and embedded systems. He is IEEE Senior Member and President of the Spanish Chapter of the IEEE Education Society.

Unai Hernández-Jayo was born in Barakaldo, Spain, in 1978. He received the M.Sc. and Ph.D. degrees in Telecommunications Engineering from the University of Deusto, Bilbao, Spain, in 2001 and 2012, respectively. In 2004, he joined the University of Deusto, where he is currently an Assistant Professor with the Telecommunications Department, teaching classes in electronic design and communications electronics. His current research interests include the use of Information and Communication Technologies in the educational process and the application of ICT in cooperative vehicular systems.

Wlodek J. Kulesza is a Professor at Blekinge Institute of Technology (BTH) in Sweden. He researches in a field of Wireless Multisensory System for Health and Security Applications. He teaches Sensors Signals and Systems as well as Research Methodology including Philosophy of Science for students of engineering. He has leaded Workshops of Scientific Writing at Chinese and Polish Universities. Among others, he is a Co-Author of the textbooks: "Scientific Metrology" and two volumes of "Measurement Data Handling".

Ingvar Gustavsson received the M.S.E.E. and Dr. Sc. degrees in Electrical Engineering from the Royal Institute of Technology (KTH), Stockholm, Sweden, in 1967 and 1974, respectively. He has been an Associate Professor of Electronics and Measurement Technology with the Blekinge Institute of Technology (BTH), Karlskrona, Sweden, since 1994. In 1999, he started a remote laboratory project at BTH that today is known as Virtual Instrument Systems in Reality (VISIR). He partly retired in 2012 to concentrate on activities related to VISIR. His research interests are in the areas of instrumentation, remote laboratories, industrial electronics, and distance learning. He has resigned from many committees, but he is still a member of Swedish professional societies.

Kristian Nilsson received the Licentiate degree from Blekinge Institute of Technology (BTH), Karlskrona, Sweden, in 2014, where he is currently working with the VISIR Open Laboratory Platform, and has been a Lecturer at BTH since 2010. His main research interests are related to distance learning, remote laboratories, electronics, and signal processing.

Johan Zackrisson is currently working as a Software Architect, but has a 15 years background at Blekinge Institute of Technology (BTH), where he was working as an engineer, mainly focusing on online engineering and remote laboratories. Together with Ingvar Gustavsson and Kristian Nilsson, he has been one of the visionaries behind VISIR platform at BTH. Most of his time nowadays is spent in industry, building telematics systems for condition-based maintenance, monitoring and support.

Andreas Pester has a degree in Mathematics and a Ph.D. in Logic and Methodology of Mathematics. He is a Professor for Mathematics and Mathematical Modeling in the Faculty of Engineering and IT of the Carinthia University of Applied Sciences, Austria, where he teaches different courses in mathematics and mathematical modeling and investigates in online simulations and online experiments, technology-enhanced learning, and machine learning for online environments. He is IEEE Member and IAOE and Treasurer of the Austrian Chapter of the IEEE Education Society.

Danilo G. Zutin is a Senior Researcher at the Carinthia University of Applied Sciences (CUAS), Villach, Austria. His research interests are in the field of remote engineering, online laboratories, Internet of Things, distributed software architectures applied to online engineering and cloud computing, particularly laboratory infrastructure as a service (LIaaS). He is author or co-author of more than 30 scientific papers published in international journals, magazines, and conferences. Most of these papers are in the field of online engineering, remote and virtual laboratories, and issues associated with their dissemination and usage.

Luis C. Schlichting is a Full Professor at the Federal Institute of Education, Science and Technology of Santa Catarina—Florianópolis—Brazil. He graduated in Electrical Engineering and has Ph.D. from the Federal University of Santa Catarina—Brazil. His main interests in research are related to remote laboratories, power electronics, and electromagnetic compatibility (EMC).

Golberi Ferreira was born in Florianópolis, SC, Brazil, in 1966. He received the M.Sc. degree in Electromagnetism in 1995, the Ph.D. degree in Electromagnetic Compatibility in 1999, both at the Federal University of Santa Catarina, Brazil, and the Posdoctorate in Engineering Education in 2017, at the University of Nottingham, UK. He joined as a Full Professor at the Federal Institute of Santa Catarina in 1990, where he was Vice-Rector from 2011 to 2016. He is now the Coordinator of the Electronic Engineering course since 2017. His current research interests include the analysis of electromagnetic compatibility systems and remote laboratories.

Daniel D. de Bona has an M.Sc. in Electrical and Computer Engineering from the Polytechnic of Porto. He works as a Technician at Federal Institute of Science, Technology and Education at Santa Catarina/Brazil, where he works in researching on electronics, remote laboratories, EMC, and embedded systems.

Fernando S. Pacheco holds an undergraduate, master's, and doctorate degree in Electrical Engineering from the Federal University of Santa Catarina. He is a Lecturer at the Federal Institute of Santa Catarina, Florianópolis, teaching for vocational and undergraduate students in the areas of signal processing, programming, and electronics, as well as coordinating research and outreach programs.

Juarez B. da Silva graduated in 1991 in Business from the Pontifical Catholic University of Rio Grande do Sul, M.Sc. in Computer Science from the Federal University of Santa Catarina (2002), and Ph.D. in Engineering and Knowledge Management from the Federal University of Santa Catarina (2007). He is currently a Lecturer at Federal University of Santa Catarina and Remote Experimentation Lab coordinator (RExLab).

João B. Alves graduated in Electrical Engineering in 1971, at the Federal University of Pará, obtained his M.Sc. from the Federal University of Santa Catarina, Brazil, in 1973, and his Ph.D. from the Federal University of Rio de Janeiro, Brazil, in 1981, both in Electrical Engineering. He partly retired from his lecturing duties at the Federal University of Santa Catarina, Brazil, where he still supervises post-graduation students. His research interests include remote laboratories, robotics, and general systems theory.

Simone Biléssimo has a bachelor's degree in Mechanic Production Engineering and M.Sc. and Ph.D. in Engineering Production. She is a Researcher at Remote Experimentation Laboratory (RexLab) and has been a Lecturer at the Federal University of Santa Catarina (UFSC) since 2012, in Araranguá/SC, Brazil. Her main research interests are related to entrepreneurship, integration of technology in education, and remote experimentation.

Ana M. Pavani holds Engineering, master's, and Doctoral degrees in Electrical Engineering. She has been working for the Pontifícia Universidade Católica do Rio de Janeiro (PUC-Rio) since 1976. She is a Member of the Board of Directors of the Networked Digital Library of Theses and Dissertations (NDLTD) and the manager of LAMBDA, a laboratory of PUC-Rio. Her research interests are ICT supported learning, digital libraries, and management of digital collections.

Delberis A. Lima holds an M.Sc. and a D.Sc. in Electrical Engineering from the Universidade Estadual Paulista (UNESP), Ilha Solteira, São Paulo, Brazil. Currently, he is Associate Professor at Pontifical Catholic University of Rio de Janeiro (PUC-Rio). His research interests include power systems planning, operations and economics, smart grid, electricity markets, and education in electrical engineering.

Guilherme Temporão has a Ph.D. in Electrical Engineering and is an Assistant Professor at Pontifícia Universidade Católica do Rio de Janeiro (PUC-Rio). His main research areas are quantum optical communications and metrology in optical components. Since 2013, he has been involved with introducing the blended learning modality and remote laboratories in several undergraduate courses.

Susana Marchisio is a Researcher at the National University of Rosario (UNR), Argentina, and has been a lecturer at UNR since 1978, where she teaches Physics of Electronic Devices. She has a Doctoral Industrial Engineering degree from ETSII/UNED, Spain. She has been Director of the Distance Education Department and of the Postgraduate Engineering School at the Faculty of Exact Sciences and Engineering in UNR. She also teaches Methodology of Research and Educational Technology at postgraduate programs. Her research fields are engineering education, distance education, teaching with technologies, and remote laboratories.

Sonia B. Concari has a Ph.D. in Physics from the Universidad Nacional de Rosario (UNR). She has been Full Professor at UNR, Universidad Tecnológica Nacional, and Universidad Nacional del Litoral, teaching Physics. She is Lecturer of Methodology of the Investigation and Epistemology of the Technology at postgraduate programs. She has been principal researcher at many projects in the fields of solar energy and remote laboratories, as well as physics and engineering education. She is also Mentor at the Pennsylvania State University.

Federico Lerro was born in Rosario, Argentina. He has a degree in Electronic Engineering at the National University of Rosario (UNR). He is Assistant Professor at the Laboratory of Physics of Electronic Devices from the Electronics Engineering career. He is the Technical Coordinator of the Remote Laboratory, which has been his main research interest for the last ten years.

Gastón S. de Arregui was born in Rosario, Argentina, in 1972. He graduated in Informatics System Engineer, and obtained a Master degree of Renewable energy in 2015. He is Researcher in remote laboratory education system and renewable energies. He is Member of Rosario National University energy laboratory. He is Postgraduate Professor of UNR and is currently pursuing an engineering Doctorate degree (Ph.D.).

Claudio Merendino was born in Rosario, Argentina in 1961 and has a degree in Electronic Engineering at the National University of Rosario (UNR) since 1987. Since 1983, he has been a Professor in the Electronic Engineering career, in the Chair of Physics of Electronic Devices. Also taught in chairs such as Digital Techniques and Computer Technology. He participates in developments of remote laboratories in the University and equipment applied to teaching in the professional field.

Miguel Plano Electronic Engineer, professor at the National University of Rosario (UNR) in the area of the science of materials and electronic devices, researcher in education of the sciences and in renewable energies where he does activities for a Masters in Energy of the same university. He participates in developments of remote laboratories linked to both fields of research.

Rubén A. Fernández has graduated in Electronic Engineer and specialist in education of superior education. He is a Full Professor in the Faculty of Exact Sciences and Technologies of the National University of Santiago del Estero, where he teaches Power Electronic and Engineering Project. He has been Director of the Academic Department of Electronics, Director of the School of Electronic Engineering and Academic Secretary of the Faculty of Exact Sciences and Technologies. He is a researcher in teaching higher education and renewable energies.

Héctor R. Paz graduated in Road Engineer, specializing in Geometric Design of Road and Transport Engineering—Formation of Human Resources in Science, Technology and Regional Development—Numerical Methods and Computer Engineering. He is Senior Professor in Philosophy and Pedagogy. Dean of the College of Sciences and Technologies of the National University of Santiago del Estero, Argentina. Outgoing President of the Federal Council of Engineering Deans of Argentina (CONFEDI). Past President of the Ibero-American Association of Institutions of Engineering Education (ASIBEI).

Mario F. Soria is an electronic engineer, graduated from the National University of Cordoba, is also a graduate in Teacher in Secondary Education Professional Technical modality. He is the Head of Practical Work in the Electronics Department of the Electronic Engineering Career. He has been in charge of the Electronics Department of the Secondary Technical School.

Mario J. Gómez graduated in electrical engineering with orientation in electronics at the National University of Tucumán, Argentina. He is an adjunct professor at the electronics 1 and electronics 2 faculty. He currently also serves as director of the electronics career. He is also a teacher in technical high school and is currently the head of the electronics department at the school.

Nival N. de Almeida holds a Ph.D. in Electrical Engineering from the Federal University of Rio de Janeiro and he is associate professor at the State University of Rio de Janeiro (UERJ). His research areas are engineering education and intelligent systems. He held the post of Rector of the UERJ from 2004 to 2007 and he was president of the Brazilian Association of Engineering Education from 2011 to 2016.

Vanderli F. de Oliveira graduated in Civil Engineering, Doctorate degree in Production Engineering and Postdoctorate in Engineering Education. He is Professor of Federal University of Juiz de Fora where he teaches Introduction and Context and Practice in Production Engineering. He is Coordinator and Researcher of the Observatory of Engineering Education, Evaluator of courses and Institutions in Brazil and in the Arcu-Sur System of Mercusul and President of the Brazilian Association of Education in Engineering (ABENGE).

María I. Pozzo is a Researcher at National Scientific and Technical Research Council and has been an Associate Professor at the National University of Rosario (UNR) since 2000, Argentina, where she teaches Field Work for Educational Scientists. She also teaches Research Methodology Postgraduates courses at different careers and universities. Her main research interests are related to teaching at the university level, ICT in higher education and interculturality. She holds a Ph.D. in Educational Sciences from UNR and did her postdoctoral stage at the Katholieke Universiteit, Leuven, Belgium with a Coimbra Group scholarship.

Elsa Dobboletta is an English Teacher and Translator graduated from Institute of Higher Education "Olga Cossettini", Rosario, where she teaches Research Methodology. Since 1984, she has been an English Lecturer at several Engineering careers at National Technological University, Regional Faculty of Rosario. She is a Member of the research team of Dr. María Isabel Pozzo at Rosario Institute of Research in Educational Sciences from the National Scientific and Technical Research Council (IRICE-CONICET).

Brenda Bertramo is Professor and graduated at the National University of Rosario (UNR) under the supervision of Dr. María Isabel Pozzo. She teaches "Professional Practice" in Higher Education Institute of Santa Fe for Technical Studies and several humanistic subjects in high schools. Her main research interests are related to higher education teaching evaluation through surveys.

Mature Learners' Participation in Higher Education and Flexible Learning Pathways: Lessons Learned from an Exploratory Experimental Research

Rogério Duarte, Ana Luísa de Oliveira Pires
and Ângela Lacerda Nobre

Abstract Higher education institutions play an important role in promoting equity and access conditions to mature learners. Such role includes the ethical commitment to facilitate learning processes, removing barriers to mature learners' entry and persistence in higher education. This paper describes the implementation of flexible learning pathways in a technology and industrial management graduate course designed for mature learners. Findings confirm that mature learners welcome flexible learning pathways and choose the pathways that better suit their needs. Despite initial academic background differences, success rates are adequate and similar for different learning pathways, showing that mature learners are capable of bridging the gaps in their academic development. Findings also show that doubts related to the impact of some learning pathways on students' academic integration are unfounded. Considering the positive results, it is concluded that flexible learning pathways, together with the widening of entry routes to higher education, promote equity and access conditions to mature learners.

Keywords Higher education · Mature learners · Equity · Flexible learning
Learning pathways

R. Duarte (✉) · A. L. de Oliveira Pires · Â. L. Nobre
Escola Superior de Tecnologia de Setúbal, Instituto Politécnico de Setúbal, Campus do IPS,
Estefanilha, 2914-761 Setúbal, Portugal
e-mail: rogerio.duarte@estsetubal.ips.pt

A. L. de Oliveira Pires
e-mail: ana-luis-pires@ese.ips.pt

Â. L. Nobre
e-mail: angela.nobre@esce.ips.pt

© Springer Nature Singapore Pte Ltd. 2018   33
M. M. Nascimento et al. (eds.), *Contributions to Higher Engineering Education*,
https://doi.org/10.1007/978-981-10-8917-6_2

1 Introduction

The last two decades have witnessed considerable changes in higher education. Boundaries between formal and non-formal education are becoming increasingly blurred, new entry routes are available, and student population in higher education institutions (HEIs) is now much more heterogeneous. Comparing the present with the recent past, higher education is now more inclusive demonstrating the process of democratization underway in the educational system and its transformation from an elite to a mass system.

An important element contributing to the transformation of higher education is lifelong learning and the notion that learning takes place on an ongoing basis and, consequently, that education should be accessible to mature and to (traditional) young learners alike. According to the European Union Council [1], lifelong learning is a cornerstone for modern economies and essential for "Europe's competitiveness in a global knowledge economy". The political commitment to lifelong learning is definitely responsible for bringing non-traditional students to the top of many HEIs' agenda and most certainly contributed to the significant increase in the number of mature learners in higher education [2]. However, such a meaningful increase also testifies for the synergistic effects caused by the combination of lifelong learning policy and actual aspirations of the many longing for an opportunity to commence higher education studies.

With the widening of entry routes, a large number of mature learners were able to access higher education, still, having access to higher education and successfully completing graduate studies are two distinct steps. Schuetze and Slowey [3] study the nature of students enrolled in higher education in several OECD countries, and based on equity considerations discuss the opportunities available for mature learners. These researchers challenge the benefit for mature learners of the changes implemented by HEIs and reflect on how HEIs integrate non-traditional students: Did these students became a part of the HEI mission? The answer to this question is crucial when discussing equity and access, because teaching and learning processes were designed for (traditional) young learners, not for mature learners that enrol after having interrupted their studies for a long period of time and that need to reconcile their academic development with professional and family responsibilities [4, 5].

If the factors that lead mature learners to participate in higher education are analysed, four categories emerge (Davies et al., 2002 in Pires [6]):

- Economic context with emphasis on labour market prospects;
- Individual circumstances—personal and social aspects, previous educational trajectory, qualifications, motivation and expectations, perception on the job market and value given to HE education, family and social support;
- Education policy—determined nationwide which include widening of entry routes, tuition fees, fiscal benefits;

- HEI policy and practice—specific to the HEI and including more flexible learning paths, evening classes, mentoring programmes as well as the academic services available to mature learners.

When mature learners initiate their studies in a HEI, they definitely give due consideration to the above-mentioned factors. However, among these factors, the fourth one, referring to the HEI policy and practice, is probably the worse known. Indeed, when entering higher education studies, mature learners are seldom aware of the pedagogic and scientific-related standards they will have to respond to, but trust the HEI to provide fair opportunities. Since HEIs have control over their teaching and learning process, they should do their best and develop the most adequate methodologies to guarantee equity for mature learners.

In this chapter, a specific methodology known as flexible learning is researched. This methodology is discussed in Kirkpatrick [7] and in Collis et al. [8] and, as the name suggests, provides students the flexibility to choose their learning pathways. According to Collis et al. [8], students' learning opportunities are improved if instead of imposing a rigid learning model, with rigid course contents, time of delivery, method of delivery and support delivery; students are allowed to choose with respect to each of these key dimensions and custom learning pathways are made available. The flexibility to choose among different learning pathways is important for mature learners in many ways; it represents the opportunity to select what, when, where, how and with whom to study. This enables individually negotiated learning activities addressing mature learners' specific needs and allows a better management of conflicts due to professional and family responsibilities. However, moving from rigid to flexible learning is difficult to put into practice. Collis et al. [8] explain that difficulties arise from cost and from conflicts for the teacher, student, student's employer and HEI while attempting to offer increased flexibility on several dimensions. Some of the dimensions described previously are questioned by both teachers and students, notably, course content flexibility. Therefore, prior to the decision to implement flexible learning, it is wise to confirm that conflicts can be managed and that students learning opportunities are actually improved.

To minimize the risk associated with changes to the teaching and learning process, a strategy is to select a pilot study, carefully monitor students' (and/or teachers') experience and research the effect of changes. For the specific case of implementing flexible learning pathways this strategy should uncover students' use of different pathways and allow the identification of the most used learning pathways. However, with flexible learning, students are free to choose their learning pathways and, because mature learners (who have full-time jobs) base their selection mostly on scheduling constraints rather than on pedagogic or scientific arguments, students' pathway selection tells little about the impact of pedagogic or scientific changes. Hence, another important objective when investigating flexible learning is to report the effect of different learning pathways on students' academic performance and integration. This is not as simple as it seems, because students' academic performance and integration also depend on students' antecedents. For example, different studies state that students' academic background (e.g. secondary education grade point average, GPA) is the best predictor of academic performance [9, 10]; age too is frequently associated with students' academic integration [9, 11]. This is the reason why data on students' antecedents (secondary education GPA; age; gender), academic performance, academic integration and the detailed tracking of each student learning pathway is needed to research the effect of flexible learning pathways.

This chapter shares the experience gathered in a process of implementing flexible learning pathways in a graduate technology and industrial management course designed for mature learners. Because a case study is included, the setting in which the research took place is addressed first. A generic characterization of mature learners is presented, and the benefit of promoting access to higher education for mature learners is discussed. Since the case study focuses on a graduate course of a Portuguese HEI, information about entry routes to Portuguese HEIs and data specific to mature learners at Instituto Politécnico de Setúbal—the researched HEI—are also presented. After defining the research setting, the implementation of flexible learning pathways and the research methodology are explained. Results for academic performance and integration considering different learning pathways are presented and, using these results, the benefits of implementing flexible learning (together with flexible entry requirements) to promote truly equitable conditions for mature learners are discussed. The chapter concludes with a review of the research results and with an appraisal of improvements and future usage of the research methodology.

2   Mature Learners in Higher Education

The complementary concepts of traditional and non-traditional student are usually related (respectively) to young learners, who enter higher education before the age of 20 in a direct transition from secondary education and enrolling in full-time programmes, and, to mature learners—also named adult students—returning to an education programme after a period of labour market activity and attending lessons while retaining their full-time jobs. A literature review shows that for several OECD countries [12], the growing number of mature learners in higher education makes the distinction between traditional and non-traditional students less meaningful; presently, a heterogeneous student population consisting of both young and mature learners is becoming the norm.

For HEIs, the contribution of mature learners is beneficial in many ways; the environment at the HEI campus benefits from a context where students of different age groups (often of different generations) socialize as peers, mature learners bring to the classroom their rich personal and professional experiences, and this diminishing of the divide between the HEI, and its surrounding community has spillover effects which largely exceed the sphere of action of higher education. Indeed, if one reflects upon how contemporary societies are organized, in particular within the setting of the information age and of the knowledge economy, providing learning opportunities to mature learners is essential for the creation of a local labour market that is more qualified and motivated to solve the problems of its region.

In regions with a strong industrial sector—the case of the district of Setúbal where the researched HEI is located—it is natural that industry and HEIs partner up to improve educational curricula and enrich applied research teams. In this context, mature learners play an important role, because, unlike traditional young learners, they are simultaneous stakeholders at HEIs and at their employer and have a direct effect on the productive system of both these institutions. With mature learners, an alliance uniting HEIs and industry is formed for which industry provides employees who become narrators of a learning process they help create.

It is this framework, which addresses the benefit of mature learners for HEIs and to the surrounding community that HEIs should consider while discussing the participation of mature learners in higher education. For a successful participation, HEIs need to take into account mature learners' schedule limitations, a larger gap in academic development, and also motivational dynamics, such as intentions, expectations and projects, which are quite different to those of young learners [13, 14].

The subject of mature learners' motivations to participate in higher education is developed in the next section, and then the specific case of mature learners in Portugal is addressed with the presentation of M23, a new entry route for Portuguese mature learners.

Mature learners' motives to participate in higher education

Mature learners participation in HE is anchored in complex motivational dynamics associated with intrinsic and extrinsic factors which evolve according to lifetime events and can be related to a wide range of causes, namely, the learning in itself (epistemic reasons), career progression, professional improvement, development of professional and personal competences, economic advantage, pressures at work, pleasure to be with others in interactive situations [6, 14].

Studies developed in Portugal centred on the M23 students confirmed the abundance of reasons and factors that trigger mature learners enrolment in higher education [14–17], including combinations between personal and professional reasons, as well as institutional and structural ones.

Based on results from a survey made in 2014/15 to 36 first-year students of a graduate course designed for mature learners—described in Sect. 3—it was found that the most common motive to enrol was the vocational (individually and combined with other motives) with 41% of the students mentioning it. Figure 1 shows this result as well as other motives mentioned during the survey.

The vocational motive is associated with the instrumentality of the course for students' career objectives, i.e. to get promoted, to find a new job. The second most mentioned motive, the professional-operational motive, is associated with the development of skills for professional projects and is mentioned by 25% of the students. A smaller number of students (12%) mentioned the identity-based motive, associated with students that enrol to improve their self-image, and the epistemic motive (11%), mentioned when students enrol for the pleasure of learning. An even

Fig. 1 Students' motives to enrol in a graduate course designed for mature learners. Inside each circle is the percentage of answers mentioning the represented motive, either individually or combined with other motives (sample: 36 first-year students, 56% of which M23 and 98% with a full-time job)

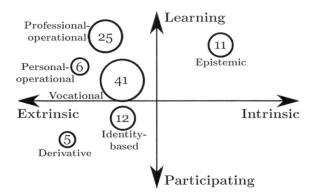

smaller number of students (6%) mentioned the personal-operational motive, associated with the development of skills for personal projects, and the derivative motive (5%), mentioned when enrolment is a way to escape from an unpleasant situation, i.e. a way out of an unpleasant professional activity.

In Fig. 1, Carré's typology of motives to participate is used and motives are presented in Carré's four axis—extrinsic versus intrinsic; learning versus participating [18, 19]. For example, the epistemic motive is intrinsic and oriented towards learning; the derivative motive is extrinsic and oriented towards participating. Four motives identified by Carré were not represented in the survey; these were the prescribed, economic, socio-affective and hedonic motives, all belonging to the "participating" axis of Carré's theoretical framework.

The fact that two thirds (66%) of the motives justifying mature learners' enrolment were related with their professional sphere (41% vocational plus 25% professional-operational) testifies for the instrumentality of higher education and, consequently, show that mature learners enrol with clear objectives: career advancement and developing skills useful to their professional activity.

Undoubtedly, mature learners decision to enrol in higher education is not triggered by the need to fulfil inner needs and even less so for the pleasure of participating. Mature learners lead busy lives and, especially in their first year, during an initial stage of academic integration, it is important that HEIs provide mature learners with the opportunities they need for successful academic performance and effective academic integration. The flexible learning methodology described in this study is an attempt to fulfil this objective.

M23: A new entry route for Portuguese mature learners

Like many other European countries, Portugal has witnessed in recent years a general improvement in the level of education and training [6]. This is a consequence of a political and socioeconomic environment favourable to an increase in the population level of education and has had the support of a lifelong learning framework responsible, among others, for the widening of entry routes to higher education.

Table 1 M23 students enrolled for the first time from 2010/11 to 2014/15

	2010/11	2011/12	2012/13	2013/14	2014/15
Total number for IPS	290	296	298	194	213
% for IPS	20	22.8%	23.6%	17.8%	18.9
% for ESTSetúbal	25	31	34.5	19.4	15.9

Source IPS Management Reports 2010, 2011, 2012, 2013, 2014 and 2015

In Portugal, a specific law was approved in the beginning of 2006 allowing access of mature learners to HEIs without the previously required degree. This entry route is now available together with the traditional route that consists of a national application after completing the secondary education. The legislation of 2006 refers to candidates with more than 23 years old—the M23 students—and takes into consideration knowledge and skills gained in other contexts of life, through professional and personal experience. The main dimensions used in the selection process are the professional and educational curriculum vitae, the candidate motivations as well as the analysis of the scientific area of study chosen. Additionally, for technology courses, a mathematics exam is required.

Considering the specific example of Instituto Politécnico de Setúbal (IPS), a medium-sized (6000 students) Portuguese public HEI located at Setúbal, an analysis of the number of M23 students enrolled for the first time between 2010/11 and 2014/15 is presented in Table 1. This table also includes percentages of M23 students for ESTSetúbal, IPS' engineering college, responsible for the researched course.

Table 1 shows that between 2010/11 and 2014/15, the percentage of M23 students enrolled for the first time varied between 18 and 24% for IPS and between 16 and 35% for ESTSetúbal. For both IPS (with a total of 5 colleges) and ESTSetúbal (the engineering college), there has been a decline in the percentage of M23 students enrolled for the first time. This decline coincides with a severe economic crisis in Portugal and highlights the need IPS had to find arguments that helped prospective mature learners make the decision to enrol and that prevented student dropout.

3 The Technology and Industrial Management Graduate Course

Technology and Industrial Management (T&IM) is a graduate course ("Licenciatura") developed at ESTSetúbal, the engineering college of IPS. This course was designed in 2006 for blue-collar workers in industrial companies located at Setúbal and nearby districts and sought to complement these workers solid technical skills (obtained from their work practice) with theoretical knowledge in business management and in engineering.

With T&IM, ESTSetúbal developed hybrid training where the engineering traditional stance was extended to account for a transdisciplinary perspective on engineering education. Transdisciplinarity as an answer to the challenges posed by societal transition, namely, from the changes brought by the evolution from an industrial to a post-industrial society for which products and services, tangible and intangible production, need novel ways to be designed, planned, produced and taken to the market.

From its conception, T&IM was believed to be relevant in the field of engineering education, not only because of innovative pedagogic and curricular approaches, but especially because T&IM targeted mature learners attending their studies while having a direct effect on the productive system of the HEI surrounding community. Besides being transdisciplinary, T&IM had other characteristics that set him apart from traditional engineering courses (for traditional young learners), namely its curriculum structure and the fact that it implemented a blended-learning (b-learning) methodology. These characteristics are detailed in the next sections.

Curricular structure

Taking into account that most mature learners have a full-time job, evening classes were scheduled two/(at the maximum) three days per week and three-course units per trimester were considered. The curriculum was designed for the course duration of four years, with total ECTS European Credit Transfer and Accumulation System, number of 180, in agreement with the Bologna Process (traditional 180 ECTS engineering courses are three-year courses).

The course curriculum was divided in equal parts between course units from management science and course units from engineering, each representing 43% of the total ECTS. The remaining 14% was divided between course units from mathematical sciences and project/internship taking place during the two last trimesters of the course. The internship was devised primarily for students who do not have a job, while working students typically address project topics related to their professional activity.

B-learning

The T&IM course implemented a b-learning methodology, blending conventional face-to-face lessons with e-learning (online autonomous learning). However, this methodology applies only to expository and problem-solving lessons not to laboratory lessons, which are always face-to-face.

The e-learning activities are synchronous or asynchronous; regardless of their type, online activities (project, chats, forum, shared work, self-test, conference–video, etc.) are designed to promote autonomous learning.

While designing the T&IM course a great deal of thought was given to the teaching and learning methodology that better suited the needs of mature learners with full-time jobs. The decision to use the b-learning methodology presented the disadvantage of less face-to-face contact hours between student, faculty and peers. According to Tinto [20], who studied traditional students and residential HEIs, student integration in the academic environment plays an important role in academic achievement and dropout; however, Bean and Metzner [9] and Tharp [21] argue that for mature learners and commuter HEIs, academic integration plays a less important role. With b-learning, students had the chance to better reconcile professional, family and academic responsibilities, having only two/(at the maximum) three days of face-to-face classes per week. This meant that students (often working shifts) could better manage the time spared from work and family and perform the required independent e-learning activities. In spite of the risk it represented, it was decided that the online component of the b-learning methodology enabled adequate autonomous learning and allowed the development of social ties between students, faculty and peers.

In a study performed in 2011/12, Lourenço et al. [22] did a survey to know T&IM students' opinion about the course curriculum. These researchers found that, overall, students were satisfied and concluded that the scheduling of curriculum activities ranked highly in students satisfaction, contributing to reconcile their academic, professional and family activities.

4 Description of the Flexible Learning Pathways

In spite of the positive feedback from students when questioned about the satisfaction with the teaching and learning process implemented in T&IM [22, 23], the decline in mature learners—mostly M23 students, see Table 1—triggered the need to implement flexible learning pathways.

In 2014–2015, flexible learning was implemented due to the availability of:

i. An extended online version of expository lessons including digital contents (e.g., videos) to catch up on face-to-face lessons.
ii. Laboratory lessons (4 h) on Saturday mornings, every two weeks.

Figure 2 presents the different learning pathways considered in 2014–2015. Bold lines in Fig. 2 describe the T&IM course traditional learning pathway, dashed lines

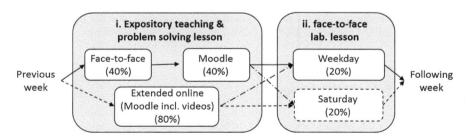

Fig. 2 Learning pathways available at the beginning of each week. The bold lines represent the traditional learning pathway. Dashed lines represent learning pathways introduced in 2014/15. Weekly course credit percentages are presented (between parentheses) for different lessons. Considering the two stages (i) expository lesson, (ii) laboratory lesson, with two alternatives each, a total of four different learning pathways are available

represent the added learning pathways. From the product of the two alternatives for each one of the two stages—(i) expository lesson and (ii) laboratory lesson—four pathways were available to students at the start of each week.

The specific circumstances that led to the decision to implement extended online lessons and Saturday laboratory lessons and a discussion of advantages/disadvantages of this decision are presented next.

Extended online lessons

Mature learners with full-time jobs sometimes miss lessons due to their professional and family responsibilities. To allow students to catch up on missed lessons, digital contents (e.g. videos) were available online for each face-to-face expository teaching and problem-solving lesson. The extended online pathway was designed to augment, not to replace face-to-face expository lessons completely, but, because attendance to face-to-face expository lessons was not mandatory, the possibility that simultaneous delivery of online contents and face-to-face lessons could "deceive" students was considered. In fact, and according to Bell et al. [24] and Dobozy [25], students lacking learner control (with trouble managing time spent studying, pace, depth and coverage of content) often believe they can use online contents to replace face-to-face lessons completely, but end up missing the contact with faculty and peers and often fail to achieve their study objectives. To detect and prevent these problems, close monitoring of the pedagogic experiment became mandatory.

Saturday laboratory lessons

A barrier to students' participation and persistence in first cycle studies is lesson scheduling conflicts. Letting students attend mandatory laboratory lessons on different dates (Saturdays or weekdays) minimizes scheduling conflicts. The financial burden and the time spent commuting to attend laboratory lessons is also reduced with Saturday lessons every two weeks instead of weekly lessons. However, since on Saturdays most academic services are closed and the number of teachers and students in the HEI campus is small, students taking Saturday lessons get a different —perhaps lessened—academic experience, which could impair their academic integration. This too suggested the need for a close monitoring of the pedagogic experiment.

The next section details the research methodology used to monitor the pedagogic experiment.

5   Research Methodology

The research took place at the start of 2014–2015. For this academic year, T&IM course had 51 students enrolled. Students' ages ranged from 18 to 48 (median: 27) and the majority (78%) were men. More than half (56%) were M23 students benefiting from lifelong learning legislation to apply and enrol, as previously described. Another important contingent were students enrolled in daytime courses that asked to be transferred; in 2014–2015, this group represented 15% of the enrolled students. The remaining students (29%) enrolled after 12 years of continued education in regular secondary schools or equivalent technological education institutes. Ninety-eight per cent of these students had a full-time job.

Out of the 51 initial students, eleven (22%) did not attend lessons and did not take any of the first trimester tests or exams. Reasons presented by these 11 students to decline the opportunity to complete the first trimester modules were: prolonged illness (3), academic credit transfer acceptance (2) and professional reasons (5). One student could not be contacted. Out of the 40 students that were assessed four reported professional difficulties and attended less than half of the laboratory lessons, the mandatory minimum attendance. The study sample considered 36 students, 70% of the initial T&IM student population.

5.1 Data Gathering

Data were gathered at the start, during and at the end of the first trimester of 2014/15. At the start of the trimester students,' sociodemographic data were collected. During the trimester, a log of lesson attendance was kept and, at the end of the trimester, data on students' academic performance and integration were collected. A brief description of the data gathering methods is presented next.

Sociodemographic data
Students' sociodemographic data including age, sex and academic background were obtained from the HEI information system. Because enrolled students came from different groups (M23, transfers from other graduate courses, secondary education), and since the rules for ranking students varied between groups, the position of each student in their group rank order was used as a measure of student's academic background. Three tiers associated with group rank order were considered.

Lesson attendance data
During the trimester, a log of lesson attendance was kept for every lesson and for each student.

Academic performance
To assess academic performance, first trimester GPA data (measured on a 20-point rating scale) were gathered from the HEI information system. Students' success was linked to GPA greater or equal to 10.

Academic integration
Students' academic integration was assessed at the end of the trimester using the QVA-r psychometric scale [26]. The QVA-r scale considers five factors of academic integration:

- *Personal*, related to students' perception of well-being.
- *Interpersonal*, related to students relationships with friends and colleagues within the HEI context but also related to the development of relationships with significant others.
- *Career*, related to students' vocational projects and also satisfied with the course.
- *Study*, related to study skills and daily study routines (e.g., time management, media used).
- *Institutional*, related to students' generic opinion about the HEI and about the academic services offered.

The QVA-r instrument was originally developed to assess academic integration in Portuguese HEIs but has also been used in other Portuguese speaking countries. The use of the QVA-r scale is reported, for example, in Almeida et al. [27] and in Igue et al. [28] for a Brazilian HEI.

5.2 Statistical Analysis

Using the log of lesson attendance, the number of students present in different types of lessons (expository, laboratory) was obtained and statistics for lesson attendance were determined. The log of lesson attendance was also used to determine, for each students, the preferred (most used) learning pathway and to populate each one of the four learning pathways represented in Fig. 2.

Considering students' academic performance and integration, students' antecedents and students' preferred learning pathways, the following hypothesis were tested:

- H_0^{Perf}: Students' academic performance is the same, regardless of the learning pathway.
- H_0^{Integ}: Students' academic integration is the same, regardless of the learning pathway.
- $H_0^{Sociodem}$: Students' sociodemographic characteristics are the same, regardless of the learning pathway.

Due to the small sample size, for continuous and ordinal variables such as age, GPA or QVA-r factors measured using the Likert scale, nonparametric Mann–Whitney tests were used to test differences between two ($i = 1, 2$) independent learning pathways. The tests considered hypothesis H_0: $F(x_1) = F(x_2)$, that variable distributions $F(x_1)$ and $F(x_2)$ were identical, against the hypothesis H_1: $F(x_1) \neq F(x_2)$, that the variable distributions were not identical. For every test performed, the Mann–Whitney U statistic and the corresponding exact two-tailed p value were determined (Version 20.0 of the IBM SPSS software [29] was used in all statistical analysis.).

For categorical variables such as gender, contingency tables were used to compare observed and expected variable counts considering two learning pathways. Fisher's exact tests were used to test if the counts were identical or not (hypothesis H_0 and H_1, respectively) and two-tailed p values of the Fisher's exact tests were determined for every hypothesis tested.

Since students' sociodemographic characteristics are an antecedent influencing academic performance and integration, together with the independent analysis of the outcome for each hypothesis tests, a joint analysis for specific combinations of outcomes is also meaningful. Of particular interest is the analysis of the outcomes for academic performance and integration given that sociodemographic characteristics of pathways are identical or different. For example, if two pathways have identical sociodemographic characteristics ($H_0^{Sociodem}$ is true), but there are differences in academic performance and integration (H_0^{Perf} and H_0^{Integ} are false) this is a sign of unbalance in pathway design. If, on the other hand, there are sociodemographic differences in pathways ($H_0^{Sociodem}$ is false), but outcomes for academic performance and integration are identical (H_0^{Perf} and H_0^{Integ} are true), this is sign that pathway design contributes to similar achievements. In this latter case, if, additionally, academic performance and integration are adequate, then, pathway design may have helped students with less favourable sociodemographic profile, for example, students who interrupted their studies for a longer period and students with more difficulties in mathematics.

6 Results

Lesson attendance and preferred learning pathways
From the log of lesson attendance, it was concluded that different students (Stdt) chose different weekly learning pathways. Table 2 presents attendance statistics for different types of lessons: face-to-face expository teaching and problem solving (Expos), laboratory on weekdays (LabWk) and laboratory on Saturdays (LabSat).

Table 2 shows that students attended most of the face-to-face expository and problem-solving lessons. The 25th percentile attendance for this type of lesson was 73%, which means that 75% of the students attended more than 73% of these lessons. On average expository and problem-solving lessons' attendance was 83%, the attendance median was 91%.

Table 2 Number of students and attendance statistics per lesson type

Lesson type	No. of Stdt (N)	Attendance statistic (%)					
		Mean	Min.	P_{25}	P_{50}	P_{75}	Max.
Expos	36	83	0	73	91	100	100
LabWk	22	82	50	70	90	98	100
LabSat	14	88	60	80	100	100	100

The large attendance percentages in face-to-face expository lessons mean that very few students relied exclusively on the extended online lessons. The adverb "exclusively" is highlighted because results do not allow the conclusion that these contents were not used, just that they were not used exclusively.

Large attendance was also registered for laboratory lessons. On average, attendance was 82 and 88% (90 and 100% medians) for weekday and Saturday laboratory lessons, respectively. During the trimester 61% (22/36) of the students preferred weekday laboratories, the remaining 39% (14/36) preferred Saturday laboratories.

Because very few students relied solely on the extended online lessons, it was decided to focus on the analysis of the availability of laboratory lessons on weekdays or on Saturdays mornings. Out of the four pathways described in Fig. 2 only the following two independent pathways were considered:

- *Weekday pathway*, comprised of a face-to-face expository lesson plus (regular) Moodle support, plus a (2 h) laboratory lesson on a weekday.
- *Saturday pathway*, similar to the above but with a (4 h) laboratory lesson on a Saturday, every two weeks.

In the following subsections, the hypothesis introduced in Sect. 5.2 is tested considering these independent pathways.

It is worth mentioning that although the number of pathways used in the statistical analysis decreased from four to just two, the result that only a small set of students chose to depend exclusively on online contents is very interesting. This result is further discussed in Sect. 7.

Academic performance

Median GPA measured on a 20-point rating scale and results from the Mann–Whitney tests are presented in Table 3 for the studied pathways. The sample median is 12.2. No statistically significant differences in GPA were found between pathways ($p > 0.10$).

Observed counts of students with GPA above 10 and results of Fisher exact tests are also presented in Table 3. Sample success rate (GPA ≥ 10) exceeded 80% (29 out of the 36 students successfully completed the trimester). Considering the 0.10 significance level, Fisher's exact tests showed no difference in success between pathways ($p > 0.10$).

From these results, it can be concluded that hypothesis H_0^{Perf} is accepted and students' performance is similar, and adequate, regardless of the learning pathway.

Academic integration

QVA-r factors' medians and results from the Mann–Whitney tests are presented in Table 3 for the studied pathways. Factors' sample medians vary between 3.23 and 3.92 on a 5-point Likert–type rating scale. All QVA-r factors have medians above 3 with factors "Career", "Personal" and "Interpersonal" having the highest sample medians. Results show that no statistically significant differences were found between pathways ($p > 0.10$ for all QVA-r factors).

Table 3 Counts, medians, Fisher's exact test results and Mann–Whitney test results for the weekday and Saturday pathways

			Pathway	
		Sample	Weekday	Saturday
Academic performance		(N = 36)	(N = 22)	(N = 14)
GPA	Median	12.2	13.3	11.5
(0–20)	Mann–Whitney U	–	126	
	p value	–	0.377	
GPA ≥ 10?	Yes count	29	19	10
(Yes/No)	No count	7	3	4
Fisher exact test p value		–	0.394	
QVA-r factor (1–5)		(N = 34)	(N = 20)	(N = 14)
Personal	Median	3.73	3.65	3.85
	Mann–Whitney U	–	119	
	p value	–	0.472	
Inter-	Median	3.65	3.73	3.58
Personal	Mann–Whitney U	–	124	
	p value	–	0.585	
Career	Median	3.92	3.85	4.00
	Mann–Whitney U	–	118	
	p value	–	0.440	
Study	Median	3.23	3.23	3.08
	Mann–Whitney U	–	118	
	p value	–	0.440	
Institutional	Median	3.62	3.56	4.00
	Mann–Whitney U	–	102	
	p value	–	0.182	
Sociodemographic characteristics				
Age	(18–48)	(N = 34)	(N = 22)	(N = 12)
	Median	28	28	38
	Mann–Whitney U	–	91.0	
	p value	–	**0.076***	
Enrol. rank position	(1–3)	(N = 34)	(N = 22)	(N = 12)
	Median	1	1	2
	Mann–Whitney U	–	57.0	
	p value	–	**0.003****	

Sample counts and medians, and the number of students (N) per pathway are also presented. All p values are two-tailed; (*) $p < 0.10$, (**) $p < 0.05$

From these results, it can be concluded that hypothesis H_0^{Integ} is accepted: students' academic integration is similar, and adequate, regardless of attending weekday or Saturday laboratory lessons.

Sociodemographic characteristics
Observed and expected counts of female and male students and Fisher's exact tests were made (results not presented). No statistically significant difference in the counts of female and male students was found for the studied pathways ($p > 0.10$).

Age and enrolment rank position medians and results from the Mann–Whitney tests are presented in Table 3. These results show that for age or enrolment rank position statistically significant differences were found between students attending weekday or Saturday pathways. Differences in age are statistically significant at the 0.10 level ($p = 0.076$), and differences in enrolment rank position are significant at the 0.05 level ($p = 0.003$).

From these results, it can be concluded that hypothesis $H_0^{Sociodem}$ is rejected: Students attending Saturday laboratory lessons are older (median is 38, whereas for the weekday laboratory is 28) and enrol with a lower rank position (median is 2, whereas for the weekday pathway is 1).

7   Discussion

One of the initial objectives of this study was to know if students used different learning pathways. Results (Table 2) show that, overall, the traditional pathway including the face–to–face expository lesson and the weekday laboratory lesson was the most used. However, pathways including Saturday laboratory lessons were preferred by almost 40% (14/36) of the students. Because students choose the pathways that better suit their needs, the fact that different pathways were used confirms that flexible learning addresses mature learners' needs.

With the availability of digital contents (e.g., videos) for each expository lesson a reduction in traditional face-to-face lessons attendance was expected. Among faculty, face-to-face lessons attendance reduction is a sensitive subject [7], and flexible learning literature presents warnings against the negative impact on students' academic integration of the simultaneous delivery of online contents and face-to-face lessons. The most common argument is that students, especially commencing students, depend on the support given by faculty during face-to-face lessons. In Bell et al. [24] and Dobozy [25], it is argued that the availability of online contents can be deceiving; especially for students lacking learner control, who trust on their ability to use the online contents to catch up or even replace face-to-face lessons completely, but end up missing the contact with faculty and peers.

Contradicting these findings, data gathered in this study do not show a reduction in face-to-face lessons attendance. Face-to-face expository lessons attendance was very high and very few students relied exclusively on the extended expository lesson pathway. These results are similar to those presented by Jones and Richardson [30] who also concluded that the delivery of online contents does not imply a reduction in face-to-face lesson attendance. According to Jones and Richardson [30], attendance depends on students' commitment to learn, regardless of the simultaneous delivery of online contents. Results presented in McShane et al. [31] show that students are not deceived by online contents; quite the opposite, students that use digital media become more concerned not to miss anything that is provided, either face-to-face or online.

The fact that the majority of the T&IM students are mature learners with full-time jobs is fundamental to explain the results from this study. For these students, academic development is perceived as instrumental for career development. This has been observed by Pires [32] and was confirmed by the results presented in Sect. 2 (see Fig. 1). In spite of the difficulties related to lower self-confidence (especially for mathematics, physics and chemistry modules included in technology courses [27, 33]), and in spite of scheduling conflicts, the perceived instrumentality of higher education studies provides mature learners the commitment needed to persist and to seek all the support available, either face-to-face or online. Although results from this study are insufficient for definite conclusions, it seems reasonable that, as mentioned by Jones and Richardson [30] and McShane et al. [31], mature learners used online contents to augment face-to-face lessons and to catch up on missed lessons, not to replace these lessons completely.

Another important topic addressed in this study is students' choice of pathway and how it relates to academic performance. Results show that students' success rate is high (exceeding 70%) regardless of the learning pathway, and hypothesis H_0^{Perf} is accepted. But rejection of hypothesis $H_0^{Sociodem}$ confirms that students' antecedents are not independent of the learning pathway chosen. Students attending Saturday laboratory lessons are older and enrol with a lower rank position (because a mathematics exam is used to rank candidates to technology graduate courses, this also means that at the beginning of the trimester students attending Saturday laboratories have lower skills in mathematics). Combining the fact that at the start of the trimester students skills are different and at the end of the trimester students' academic performance is similar and adequate, it can be concluded that the adopted teaching and learning process helped students overcome the gap in their academic development process.

Finally, this study addressed the link between students' chosen learning pathway and students' academic integration. Results show that hypothesis H_0^{Integ} is accepted and academic integration does not vary with the learning pathway. Furthermore, for the Saturday pathway academic integration results are adequate (QVA-r factors above 3.00), dismissing initial doubts related to the negative impact of having lessons on Saturdays.

In summary, despite initial sociodemographic differences between weekday and Saturday laboratory pathways, choosing either one results in identical and adequate academic performance and integration.

8 Conclusion

Flexible learning pathways were implemented in a technology and industrial management graduate course designed for mature learners. Results show that students choose the learning pathways that better suit their needs, and that success rates are similar and adequate (exceeding 70%) regardless of the learning pathway.

Because different pathways were chosen by students with different characteristics—notably, with different academic backgrounds— similar and adequate success rates are evidence that gaps in students' academic development process were successfully bridged. Moreover, results also show that initial doubts related to the negative impact of Saturday laboratory lessons (that could provide a lessened academic experience) and that initial doubts related to expanding online contents (that could lead students to think expository and problem-solving face-to-face lessons were dispensable) were unfounded. Students attending laboratories on Saturdays do not rate their academic integration as inferior and students still attend face-to-face expository and problem-solving lessons after extending the online contents available.

In spite of the results supporting the use of flexible learning, the researched implementation was ineffective for students that attended less than 50% of the laboratory lessons and for the 20% that failed. For these students, dimensions of flexible learning not included in the present study (e.g.,time, pace and course content flexibility) could provide the extra support needed. Also, in future studies, some methodological improvements are worth considering, notably, mature learner characterization would benefit from information on the initial ability to perform autonomous work and on basic skills necessary to use e-learning tools. Moreover, alongside the monitoring of academic performance and integration, it would be interesting to observe the changes to students' motivation patterns as they advance in their studies.

Even if the conclusions from this study apply to a specific group—mature learners—and to a specific technological course, the research methodology is generic. With the support of HEIs information systems and developments in academic analytics it is now much simpler to track the activity of students, learn their preferred learning pathways and their use of distant learning tools. This chapter shows the research methodology was useful to investigate the effect of flexible learning pathways for the technology and industrial management graduate course of ESTSetúbal-IPS, and this methodology could also be useful for other HEIs wishing to test changes to their teaching and learning process.

References

1. European Union Council (2011) Council conclusions on the role of education and training in the implementation of the 'Europe 2020' strategy. Off J Eur Union (2011/C 70/01) (E.U.)
2. OECD (2015) OECD statistic extracts database. http://stats.oecd.org/Index.aspx? DatasetCode=RENRLAGE. Accessed June 2015
3. Schuetze H, Slowey M (2000) Higher education and lifelong learners. International perspectives on change. Routledge Falmer, London
4. Jarvis P (1987) Meaningful and meaningless experience: towards an analysis of learning from life. Adult Educ Q 37:164–172
5. Kenner C, Weinerman J (2011) Adult learning theory: applications to non-traditional college students. J Coll Reading Learn 41:87–96
6. Pires A (2009) Novos públicos no Ensino Superior em Portugal. Contributos para uma problematização. In: Rummert, Canário, Frigotto (org) Políticas de Formação de jovens e adultos no Brasil e em Portugal. Ed. UFF, Rio de Janeiro, pp 185–205
7. Kirkpatrick D (1997) Becoming flexible: contested territory. Stud Contin Educ 19:160–173
8. Collis B, Vingerhoets J, Moonen J (1997) Flexibility as a key construct in European training: experiences from the telescopia project. Br J Educ Technol 28:199–217
9. Bean J, Metzner B (1985) A conceptual model of nontraditional undergraduate student attrition. Rev Educ Res 55:485–540
10. Cross K (1981) Adults as learners. Jossey-Bass, San Francisco
11. Lynch J, Bishop-Clark C (1994) The influence of age in college classrooms: some new evidence. Community Coll Rev 22:3–12
12. Pires A (2016) Between challenges and trends of lifelong learning: higher Education and the recognition of prior experiential learning. In: Jongbloed B, Vossensteyn H (eds) Access and expansion post-massification. opportunities and barriers to further growth in higher education particpation. Routledge, London
13. Griffiths V, Kaldi S, Pires A (2008) Adult learners and entry to higher education: motivation, prior experience and entry requirements. In: Munoz M, Jalinek I, Ferreira F (eds) Proceedings of the international association of scientific knowledge teaching and learning conference, 26–28 May 2008, Aveiro, Portugal. International association of scientific knowledge teaching and learning, Aveiro, pp 632–640
14. Pires A (2009) Higher education and adult motivation towards lifelong learning. Eur J vocat train 46:130–150
15. Pereira E (2009) Alunos maiores de 23 anos: Motivações para o ingresso no ensino superior na Universidade do Porto. Dissertação de Mestrado em Sociologia, Faculdade de Letras da Universidade do Porto
16. Fragoso A, Valadas S (2013) Older mature students in higher education. In: Paper presented at ESREA—4th international conference learning ppportunities for older adults: forms, providers and policies, 23–25 October 2013, Mykolas Romeris University, Ateitiessstr. 20 Vilnius, Lithuania
17. Quintas H, Fragoso A, Bago J, Gonçalves, Ribeiro T, Monteiro M, Fragoso A, Santos L, Fonseca M (2014) Estudantes adultos no Ensino Superior: O que os motiva e o que os desafia no regresso à vida académica. Revista Portuguesa de Educação, 27(2):33–56
18. Carré P (1998) Motifs et dynamiques d'engagement en formation. Synthèse d'une étude qualitative de validation auprès de 61 adults en formation professionnelle continue. Education Permanente 136:119–131
19. Carré P (2000) Motivation in adult education: from engagement to performance. In: Proceedings of the 41st annual adult education research conference, Canada
20. Tinto V (1975) Dropout from higher education: a theoretical synthesis of recent research. Rev Educ Res 45(1):279–294
21. Tharp J (1998) Predicting persistence of urban commuter campus students utilizing student background characteristics from enrolment data. Community Coll J Res Pract 22(3):279–294

22. Lourenço R, Ferreira E, Duarte R, Gonçalves H, Duarte J (2013) IPS' technology and industrial management graduate course: a curriculum follow-up analysis. In: Morgado J, Alves M, Viana I, Ferreira C, Seabra F, Hattum-Janssen N, Pacheco J (eds) European conference on curriculum studies, future directions: uncertainty and possibility, University of Minho, pp 263–269
23. Duarte R, Ramos-Pires A, Gonçalves H (2014) Identifying art-risk students in higher education. Total Qual Manage Bus Excell 25:944–995
24. Bell T, Cockburn A, McKenzie B, Vargo J (2001) Flexible delivery damaging to learning? Lessons from the Canterbury digital lectures project. University of Cantebury. http://ir. canterbury.ac.nz/bitstream/10092/517/1/42637_edmedia.pdf. Accessed June 2015
25. Dobozy E (2008) 'Plan your study around your life, not the other way around': How are semi-engaged students coping with flexible access. In: Proceedings of EDU-COM conference
26. Almeida L, Soares P (2002) Questionário de Vivências Acadêmicas (QVAr): Avaliação do ajustamento dos estudantes universitários. Avaliação Psicológica 2:81–93
27. Almeida L, Soares AP, Guisande M, Paisana J (2007) Rendimento académico no ensino superior: Estudo com alunos do 1º ano. Revista Galego-Portuguesa de Psicoloxia e Educación 14:207–220
28. Igue E, Bariani I, Milanesi P (2008) Vivência acadêmica e expectativas de universitários ingressantes e concluintes. Psico-USF 13:155–164
29. Arbuckle J (2011) IBM SPSS (version 20.0) [Computer Program]. SPSS, Chicago
30. Jones S, Richardson J (2002) Designing an IT-augmented student-centred learning environment. In: Proceedings of HERDSA conference. Perth, Australia, pp 376–383
31. McShane K, Peat M, Masters A (2007) Playing it safe? Students' study preferences in a flexible chemistry module. Aust J Educ Chem 67:24–30
32. Pires A (2010) Aprendizagem ao Longo da Vida, Ensino Superior e novos públicos: uma perspectiva internacional. In: Aprendizagem ao Longo da Vida e Políticas Educativas Europeias: Tensões e ambiguidades nos discursos e nas práticas de Estados, Instituições e Indivíduos. UIED, Colecção Educação e Desenvolvimento, FCT / Universidade Nova de Lisboa, pp 107–137
33. Jameson M, Fusco B (2014) Math anxiety, math self-concept, and math self-efficacy in adult learners compared to traditional undergraduate students. Adult Educ Q 64:306–322

Rogério Duarte holds a Ph.D. in Mechanical Engineering from Lisbon University, Portugal, and has more than 15 years of professional experience as a researcher, consultant, and teacher. He is a Professor in the Mechanical Engineering Department at Escola Superior de Tecnologia de Setúbal, Instuteto Politécnico de Setúbal, Portugal, and along his core research activities in building physics/energy conservation, he is also interested in promoting and researching access and equity for mature learners in higher education. He is a member of the Portuguese Engineers Society (OE) and of the Portuguese Society for Engineering Education (SPEE).

Ana Luísa de Oliveira Pires is a Professor at the Escola Superior de Educação, Instituto Politécnico de Setúbal, since 2007, in the Department of Social Sciences and Pedagogy. She is a researcher at UIED, FCT-UNL, and her main research interests are in the domain of lifelong learning, adult learning and higher education policies. She has published several articles, chapters and books at national and international level (Escola Superior de Educação, Instituto Politécnico de Setúbal, Campus do IPS, Estefanilha, 2914-504 Setúbal, Portugal).

Ângela Lacerda Nobre born in Lisbon, in 1960, has an academic background in nursing and economics, has a master in Applied Economics, a post-graduation in Philosophy, Psychoanalytical Training and a Ph.D. in Information Systems (Semiotic Learning: a conceptual framework to facilitate learning in knowledge-intensive organisations). She has professional experience as a nurse and as an economist; she has been working at a management school (esce.ips.pt) since 1998 and has published her research in the areas of semiotics, psychoanalysis, practical philosophy, knowledge management and organisational learning; the study of effectiveness, from a telocentric perspective, and its relationship to human individuation, is her kernel research interest (e.g. see YouTube).

The Flow of Knowledge and Level of Satisfaction in Engineering Courses Based on Students' Perceptions

Celina P. Leão, Filomena Soares, Anabela Guedes,
M. Teresa Sena Esteves, Gustavo R. Alves, Isabel M. Brás Pereira,
Romeu Hausmann and Clovis António Petry

Abstract In this chapter, the results of a questionnaire are analyzed to assess engineering students' satisfaction toward their courses, working conditions, and academic environment, as well as the flow of knowledge perception along the first

This work has been supported by research centre CIETI (Centro de inovação em engenharia e tecnologia industrial) in the scope of UID/EQU/00305/2013 and FCT (Fundação para a ciência e tecnologia) in the scope of the projects COMPETE: POCI-01-0145-FEDER-007043 and UID/CEC/00319/2013

C. P. Leão · F. Soares
Centro ALGORITMI, School of Engineering, University of Minho, Campus de Azurém,
4800-058 Guimarães, Portugal
e-mail: cpl@dps.uminho.pt

F. Soares
e-mail: fsoares@dei.uminho.pt

A. Guedes · M. T. Sena Esteves · G. R. Alves · I. M. Brás Pereira (✉)
Centro de Inovação em Engenharia e Tecnologia Industrial, Instituto Superior de Engenharia
do Porto (ISEP), 4200-072 Porto, Portugal
e-mail: imp@isep.ipp.pt

A. Guedes
e-mail: afg@isep.ipp.pt

M. T. Sena Esteves
e-mail: mte@isep.ipp.pt

G. R. Alves
e-mail: gca@isep.ipp.pt

R. Hausmann
Department of Electrical and Telecommunication Engineering, University of Blumenau,
Blumenau, Brazil
e-mail: romeuh@furb.br

C. A. Petry
Federal Institute of Santa Catarina, Florianópolis, Brazil
e-mail: petry@ifsc.edu.br

© Springer Nature Singapore Pte Ltd. 2018
M. M. Nascimento et al. (eds.), *Contributions to Higher Engineering Education*,
https://doi.org/10.1007/978-981-10-8917-6_3

three curricular years. With a sample of 654 students from four higher education institutions and two countries, the study focused in eleven items, concerning teachers' involvement perception, student–teacher interaction, course organization, and functioning, and overall satisfaction. Several research hypotheses were considered, and significant correlations were investigated. Results show that students, in average, are satisfied with the course and with student–teacher interaction, but perceive that teachers do not contextualize the contents in a professional perspective. The flow of knowledge is neither clearly understood. Two positive significant correlations exist between: students' overall satisfaction and their expectations; the way students assess their interaction with teachers and the way they assess teachers' involvement. No significant differences were found between the two countries.

Keywords Students' perceptions · Satisfaction questionnaire · Engineering courses · Higher education

1   Introduction

Students' satisfaction and perceptions regarding their university courses are important measures for higher education institutions. Several studies can be found in the literature with different focuses but where the common issue is to analyze students' feedback [1–12].

In particular, in Sinclaire's work [1], it is referred that often, in the literature, students' satisfaction is related to institutional concern for the quality of courses and programs and the need to understand student perceptions. This author [1] presented a study in a business course of a public university in the southeast region of the USA with 560 students by using a questionnaire delivered via the Internet (a 5-point Likert-type scale was used, where 1 meant unimportant, and 5 represented very important). The items under evaluation were institutional factors, specifically "How important are college facilities and services to student satisfaction with a college course?"; learning environment, in detail class size, time, frequency, methods of instruction, instructor characteristics and behavior, learning technology, methods of grading, and course content; and considering all the above factors, "How important is each item to your overall satisfaction with a college course?". The results pointed out that factors relating to instructor characteristics (the teacher is always available, has a working knowledge of the subject, is interested in student learning, and he/she is passionate about the subject), methods of instruction and methods of grading were considered important or very important for student satisfaction with a college course. Additionally, 86% of students rated student-oriented course factors (student interest in subject, perception that course subject applies to work or profession, course in student major) as important or very important.

Tasirin et al. [2] present a study about students' satisfaction in master engineering programs (Universiti Kebangsaan Malaysia). By means of a survey that uses the service–product concept and analyzes expectations and perceptions, students' satisfaction is evaluated in different groups of factors: non-academic aspects, academic aspects, program issues, design, delivery and assessment, reputation, and access. The overall satisfaction is then related to each of the referred issues in order to identify the aspects that may be improved in terms of service quality. Results show that the overall satisfaction relies mostly in university reputation and that the worst assessment is given to non-academic aspects (such as administrative services). Results also show that students recognize the effect of the courses in developing certain employability skills. The study also allows identifying specific topics in the learning methodologies that might need some modifications in order to improve students' competencies in professional domains and science and engineering knowledge.

Cronje and Coll [3] presented a study developed in New Zealand university and polytechnic sectors in science and engineering courses. Students' beliefs of their higher education experiences were addressed as well as how they see the learning process and if it serves their careers' aims. One of the conclusions of the study was that polytechnic students were more positive regarding their learning, where its practical component may lead to enhance career prospects.

Borges et al. [4] focused their study on business courses (management, accounting, and tourism business) in two Brazilian universities, one public and one private. A quantitative and descriptive study was performed based on 513 students' responses. The results pointed out that students have more confidence in private universities than in public institutions. Nevertheless, the authors concluded that students' trust decreases with time. Also, it was verified that female students rely more on their universities than their male mates.

Some attempts have been made in order to study cultural and social factors that may influence students' degree of satisfaction. Demographics characteristics, ethnicity, and gender were analyzed by Lord et al. [5]. This study considered a large and diverse dataset with 90,000 first-time-in-college and 26,000 transfer students who attended engineering institutions at the USA, including students who started in first-year engineering programs, those changing majors, and those transferring from other institutions.

The study presented by Vaz et al. [6] aimed to analyze the applicability of a model of student satisfaction within higher education in two different countries: Portugal and Uruguay. The index of higher education student satisfaction developed by Alves and Raposo [7] was taken as a basis and compared with Uruguay. The results showed different performances in the two countries.

Several other studies are available in the scientific community. In the present work, it is not the authors' goal to present an extensive review on students' perceptions regarding engineering courses. However, there is still scope for identifying and exploiting many other different approaches. Following this trend, this paper is focused on evaluating the students' perception in the first three years of six engineering courses in electrical/electronics area in universities and polytechnic institutions in two countries, Portugal and Brazil. By doing this, it would be possible to follow how this perception and students' satisfaction evolve over the three years of the programmes. Also, it may allow to perceive if students understand that as the years pass, the state of knowledge increases and somewhat concurrently. This is what the flow of knowledge stands for in the present work.

This chapter is organized in five sections besides the Introduction. Section two, Frame of Study, presents the main objectives of the study, and the higher education institutions and courses analyzed. Section three, Materials and Methods, details the research instrument (questionnaire used), the methodology followed as well as the sample characterization. Section four describes the research hypotheses, and the results are presented and analyzed in section five. Conclusions are drawn in the last section of the paper.

2 Frame of Study

2.1 Main Objectives

The general objective of this full-scale study is to analyze and determine the level of satisfaction of electrical/electronic engineering students toward their courses, working conditions and academic environment, as well as to assess the way students perceive the flow of knowledge along the first three curricular years and in four different higher education institutions (HEIs), from two different countries.

2.2 Higher Education Institutions and Courses Analyzed

This study considers four HEI: two from Portugal (Instituto Superior de Engenharia do Porto, ISEP, and Escola de Engenharia da Universidade do Minho, EEUM) and two from Brazil (Instituto Federal de Santa Catarina, IFSC, and Universidade Regional de Blumenau, FURB).

Six different studies' programmes were chosen: two from ISEP (electrical and computer engineering, ECE-ISEP, and electrical engineering—power systems, EE-PS-ISEP), two from IFSC (electrical engineering, EE-FSC and electronics engineering, EiE-IFSC), one from FURB (electrical engineering, EE-FURB) and one from EEUM (industrial electronics engineering and computers, IEEC-UM) [13].

For each of the six engineering degrees in electrical/electronic, the first three years will be analyzed, corresponding to the first cycle of higher education.

3 Materials and Methods

This section presents briefly the instrument used for the data collection, the methodology followed as well as the sample characterization.

3.1 Description of the Research Instrument (Questionnaire)

A questionnaire was used as the research instrument in order to fulfill the objective and validate the set of hypotheses (defined in Sect. 4). In summary, the questionnaire includes two main parts: (1) student's characterization (age, gender, higher education institution, curricular year, semester, regime, number of registrations in the course, and regular/working student), (2) a list of forty-four items, divided in six groups [teachers' involvement perception (TIP), student interest (SI), student–teacher interaction (STI), course organization and functioning (COF), infrastructures (IS), and overall satisfaction (OS)], classified in a 5-point agreement Likert scale, 1 (Strongly Disagree) to 5 (Strongly Agree), with the neutral point (3) being neither disagree nor agree. The questionnaire reliability and validity had been previously assessed [13, 14]. Other details about the questionnaire and its previous validation may be found elsewhere [13, 14]. To fulfill the purpose of this study, eleven items from four of the six groups mentioned above were analyzed. Since the questionnaire was originally written in Portuguese (mother language in both countries), the translation of the items into English will be presented in the paper expressing the same meaning as in the original language.

The eleven items under analysis in the present study are included in groups: teachers involvement perception (TIP), student–teacher interaction (STI), course organization and functioning (COF), and overall satisfaction (OS), as follows:

TIP_4: *In general, teachers aim to contextualize the contents in a professional perspective*;
TIP_6: *In general, teachers present challenges to be solved outside the classroom*;
TIP_7: *In general, students assess positively teachers' performance*;
STI_6: *In general, the student–teacher interaction is positive*;
COF_8: *Teachers try to relate the syllabus with other courses*;
COF_9: *The syllabuses of the courses are well articulated with previously acquired knowledge*;
COF_10: *In general, the courses meet my expectations*;
COF_11: *The engineering course is well organized*;
OS_1: *I'm satisfied with the school environment and working conditions*;
OS_2: *I'm satisfied with the academic environment (cultural, sporting, and recreational activities)*;
OS_3: *In general, I'm satisfied with the course.*

3.2 Methodology

The questionnaires were approved by the courses' directors and then distributed according to the availability of teachers and students timetable. This procedure was chosen not to interfere with the proper functioning of the classes, in the second semester of 2014/15 and in order to ensure a larger sample, were also distributed in the first semester of 2015/16 (to different students). A curricular unit of each curricular year was chosen to hand out the questionnaire. After an explanation of the questionnaire aim, students answered it on a voluntary basis. Filling out the questionnaire took no more than 15 min.

3.3 Sample Characterization

Out of 661 questionnaires received, 654 answered completely all items and considered valid for analysis, with 18.9% from EEUM, 26.0% from FURB, 32.6% from IFSC, and 22.5% from ISEP. Based on the purpose of the present study, this sample size was considered acceptable and adequate [15]. With a confidence level of 95% and ±3% precision, considering a correction for finite population ($N \approx 950$ students), it would be necessary a sample size of 503 students (lower than the received valid 654 questionnaires). Also, the dimensions of the four subgroups (from the four HEIs) are considered acceptable.

The mean age is 21.7 years (SD = 4.9, range 17–55 years), and most of the students (67.2%) are aged 21 years or less. In all HEI, the majority of students (86.8%) are male (EEUM 91.2%, FURB 93.0%, IFSC 74.8%, ISEP 93.3%). Regarding classes' regime, FURB and ISEP are the two HEIs that have both day and after work classes. FURB has 87.1% and ISEP has 19.3% of students in after work classes. However, all of the HEI have students with student worker status (EEUM 7.2%, FURB 85.4%, IFSC 29.4%, ISEP 18.2%).

Table 1 summarizes the descriptive statistics regarding these variables.

Table 1 Summary of the descriptive statistics of the students' characteristics (see Sect. 2.2 for acronyms)

HEI	EEUM	FURB	IFSC	ISEP	Total
Characteristics	Percentage (%)				
Questionnaires	18.9	26.0	32.6	22.5	100
Gender					
Male	91.2	93.0	74.8	93.3	86.8
Female	8.8	7.0	25.2	6.7	13.2
Age					
≤18	17.6	25.1	32.1	1.3	20.6
19 < > 21	62.4	33.9	43.3	52.7	46.6
22 < > 24	11.2	18.1	12.6	21.4	15.7
≥25	8.8	22.9	12.0	24.6	16.7
Regime of class					
Day	100	12.9	100	80.7	73.0
After work	0.0	87.1	0.0	19.3	27.0
Students with student worker status	7.2	85.4	29.4	18.2	37.2
Characteristics	Mean ± S.D.				
Age	20.9 ± 4.2	22.2 ± 5.2	20.6 ± 3.9	23.3 ± 5.7	21.7 ± 4.9

4   Research Hypotheses

In order to achieve the proposed main objective of this study more easily, sixteen hypotheses were formulated to investigate students' perceptions in the four HEIs and their evolution along the first three curricular years. These hypotheses were stated on the basis of expected students' answers to the questionnaire distributed. They are as follows:

Hypothesis 1. Students perceived positively teachers' interest in contextualizing the contents in a professional perspective.
Hypothesis 2. Students' perception about whether teachers contextualize the contents in a professional perspective is independent of the curricular year and shows a similar behavior regardless of institution.

These two hypotheses allow understanding if and how students recognize the concern of the teacher in contextualizing the contents in a professional perspective. It is known that the first year usually consists of introductory courses, in the second year some of the subjects may be more theoretical than practical, and the third year has a different character from that of the earlier years with more specific and more practical subjects. Also, it will be possible to understand if this awareness changes over the years and if it is different or similar in the four institutions.

To do this analysis, the students' answers to the item "*In general, the teachers aim to contextualize the contents in a professional perspective*" (from TIP group, TIP_4) will be considered.

Hypothesis 3. Students perceived positively teachers' interest to present challenges to be solved outside the classroom.
Hypothesis 4. Students' perception about whether teachers present challenges to be solved outside the classroom is independent of the curricular year and shows a similar behavior regardless of institution.

In a way, these two hypotheses follow the idea of the previous ones: "students perceive that teachers' challenges together with the demonstration of the syllabus applicability to real problems promote students' interaction, satisfaction, and motivation."

To do this analysis, the students' answers to the item "*In general, the teachers present challenges to be solved outside the classroom*" (from TIP group, TIP_6) will be considered.

Hypothesis 5. Students assess positively teachers' performance.
Hypothesis 6. Students' assessment on teachers' performance is independent of the curricular year and shows a similar behavior regardless of institution.

To do this analysis, the students' answers to the item "*In general, students assess positively teachers' performance*" (from TIP group, TIP_7) will be considered.

Hypothesis 7. Students assess positively student–teacher interaction.
Hypothesis 8. Students' assessment on student–teacher interaction is independent of the curricular year and shows a similar behavior regardless of institution.

If students demonstrate a positive assessment regarding this issue, this could result in a more proactive attitude toward learning.

To do this analysis, the students' answers to the item "*In general, the student–teacher interaction is positive*" (from STI group, STI_6) will be considered.

Hypothesis 9. Students are able to realize the connection between the different courses (or curricular units, CU) and contents articulation.
Hypothesis 10. Students' perception on the connection between the different courses and contents articulation is independent of the curricular year and shows a similar behavior regardless of institution.

Any engineering degree incorporates a wide range of topics/subjects (from mathematics, statistics, and management to specific electronics subjects), so it is essential to create links between topics that apparently have no connection, in a well-structured manner. One way of doing this is to include these topics in integrating projects. So, it is important to understand if students are aware of this course structure.

To do this analysis, the students' answers to the two items "*Teachers seek to relate the contents with other CUs*" and "*The CU syllabus is well articulated with prior knowledge acquired*" (from COF group, COF_8 and COF_9, respectively) will be considered.

Hypothesis 11. Students positively assess course organization and this meets the students' best expectations.
Hypothesis 12. Students' assessment on course organization and students' best expectations fulfillment are independent of the curricular year and show a similar behavior regardless of institution.

To do this analysis, the students' answers to the two items "*In general, the courses meet my expectations*" and "*The course is well organized*" (from COF group, COF_10 and COF_11) students' answers will be considered.

Hypothesis 13. Students are satisfied with the school and academic environment.
Hypothesis 14. Students' satisfaction with the school and academic environment is independent of the curricular year and shows a similar behavior regardless of institution.

Satisfied students create a good work environment, and so they can stay longer at school not only for academic work but also for sports and cultural activities.

To do this analysis, the students' answers to the three items "*I'm satisfied with the school environment and working conditions*", "*I'm satisfied with the academic environment (cultural, sporting, and recreational activities),*" and "*Overall, I'm satisfied with the course*" (from OS group, OS_1, OS_2, and OS_3, respectively) will be considered.

Besides analyzing the referred answers, some correlations were tested aiming to identify tendencies among students' perceptions, and those correlations are described by hypotheses 15 and 16.

Hypothesis 15. The student–teacher interaction assessment is positively and significantly related to how students assess teachers' performance.

If students assess their interaction with teachers positively, it would be interesting to observe if and how this is related with their teachers' performance assessment. To do this analysis, the students' answers given to items STI_6 and TIP_7 will be considered.

Hypothesis 16. There is a significant and positive relationship between the students' overall satisfaction and the students' expectations regarding the course.

To do this analysis, students' answers to items OS_3 and COF_10 will be considered.

5 Results Analysis and Discussion

The results analysis of the students' awareness of the flow of knowledge and level of satisfaction throughout the first three years of the engineering course are presented and discussed below. The SPSS statistical tool was used for the data analysis [16].

As the data collected from the students do not follow the normality (normality was checked with the Shapiro–Wilk test), nonparametric tests were considered on the analysis in accordance with the various hypotheses to test (Kruskal-Wallis (H), for the comparison of more than two independent samples as alternative to the *t*-test, Friedman (F), alternative to the *t*-test for dependent samples, and Spearman's correlation coefficient (r_S) to study the relationship between two items). A significance level of 5% was considered.

5.1 Global Results

In Table 2, the results of the eleven items are summarized in terms of mean value, standard deviation (s.d.), and median, for the three curricular years and for the four

HEI. Regarding the different courses, they were not considered separately as they come from a similar area of knowledge: electrical/electronic engineering.

The descriptive analysis of the results presented in Table 2 shows that, in average, students' assessment is positive (mean value higher than 3). More than

Table 2 Summary of the descriptive statistics of the results for the eleven items (see Sects. 2.2 and 3.1 for acronyms)

Items	Year\HEI	Mean (s.d.); Median				
		EEUM	FURB	IFSC	ISEP	Total
TIP_4	First	3.8 (0.8); 4	3.3 (0.9); 3	4.1 (0.7); 4	3.4 (0.9); 3	3.6 (0.9); 4
	Second	2.9 (0.9); 3	3.5 (0.8); 3	3.3 (1.0); 3	3.6 (0.7); 4	3.3 (0.9); 3
	Third	3.0 (0.8); 3	3.3 (1.0); 3	3.6 (1.0); 4	3.2 (0.9); 3	3.2 (0.9); 3
	Total	3.2 (0.9); 3	3.3 (0.9); 3	3.8 (1.0); 4	3.4 (0.9); 3	3.4 (0.9); 3
TIP_6	First	3.6 (0.7); 4	3.8 (1.0); 4	4.1 (0.9); 4	3.7 (0.8); 4	3.8 (0.9); 4
	Second	3.3 (1.0); 3	3.8 (1.0); 4	3.7 (1.0); 4	3.5 (0.7); 3	3.6 (1.0); 4
	Third year	3.5 (0.7); 3.5	3.5 (0.9); 4	3.7 (1.0); 4	3.2 (0.9); 3	3.5 (0.9); 4
	Total	3.5 (0.8); 4	3.7 (1.0); 4	3.9 (1.0); 4	3.5 (0.8); 4	3.7 (0.9); 4
TIP_7	First	3.7 (0.7); 4	3.9 (0.8); 4	4.3 (0.7); 4	3.8 (0.8); 4	4.0 (0.8); 4
	Second	3.2 (0.8); 3	3.9 (0.5); 4	4.0 (0.8); 4	3.7 (0.5); 4	3.8 (0.7); 4
	Third	3.4 (0.7); 3	3.8 (0.6); 4	4.0 (1.0); 4	3.5 (0.8); 4	3.6 (0.8); 4
	Total	3.4 (0.7); 4	3.8 (0.7); 4	4.2 (0.7); 4	3.6 (0.8); 4	3.8 (0.8); 4
STI_6	First	3.8 (0.6); 4	3.9 (0.8); 4	4.2 (0.6); 4	3.7 (0.7); 4	4.0 (0.7); 4
	Second	3.7 (0.6); 4	3.9 (0.8); 4	4.2 (0.8); 4	3.7 (0.6); 4	3.9 (0.8); 4
	Third	3.5 (0.7); 4	3.6 (0.7); 4	4.5 (0.7); 5	3.5 (0.8); 4	3.6 (0.8); 4
	Total	3.6 (0.7); 4	3.8 (0.8); 4	4.2 (0.7); 4	3.7 (0.7); 4	3.9 (0.8); 4
COF_8	First	3.7 (0.8); 4	3.5 (0.9); 4	3.9 (0.8); 4	3.7 (0.8); 4	3.7 (0.8); 4
	Second	3.2 (0.8); 3	3.7 (1.0); 4	3.4 (1.0); 3.5	3.7 (0.6); 4	3.5 (0.9); 4
	Third	3.6 (0.7); 4	3.6 (0.9); 4	4.0 (0.9); 4	3.4 (0.7); 3	3.6 (0.8); 4
	Total	3.5 (0.7); 4	3.6 (0.9); 4	3.7 (0.9); 4	3.6 (0.7); 4	3.6 (0.9); 4
COF_9	First	3.7 (0.7); 4	3.6 (0.8); 4	3.8 (0.8); 4	3.2 (0.8); 3	3.6 (0.8); 4
	Second	3.4 (0.6); 3	3.7 (0.8); 4	3.5 (1.0); 4	3.6 (0.7); 4	3.5 (0.9); 4
	Third	3.4 (0.9); 4	3.5 (0.8); 3	3.7 (0.8); 4	3.3 (0.9); 3	3.5 (0.9); 4
	Total	3.5 (0.8); 4	3.6 (0.8); 4	3.6 (0.9); 4	3.3 (0.8); 3	3.5 (0.9); 4
COF_10	First	3.8 (0.7); 4	3.8 (0.8); 4	4.1 (0.8); 4	3.5 (0.7); 4	3.8 (0.8); 4
	Second	3.6 (0.6); 4	3.7 (0.7); 4	3.8 (0.9); 4	3.8 (0.5); 4	3.7 (0.7); 4
	Third	3.4 (0.8); 3	3.6 (0.7); 4	4.1 (0.9); 4	3.3 (0.6); 3	3.5 (0.8); 4
	Total	3.5 (0.7); 4	3.7 (0.8); 4	3.9 (0.8); 4	3.5 (0.7); 4	3.7 (0.8); 4
COF_11	First	4.0 (0.7); 4	4.0 (0.9); 4	3.9 (0.9); 4	3.4 (0.9); 4	3.8 (0.9); 4
	Second	3.2 (0.9); 3	3.7 (0.7); 4	3.2 (1.0); 3	3.6 (0.5); 4	3.4 (0.9); 4
	Third	3.0 (0.9); 3	3.5 (0.8); 4	3.8 (0.9); 4	3.1 (0.7); 3	3.3 (0.9); 3
	Total	3.3 (1.0); 3.5	3.7 (0.9); 4	3.7 (1.0); 4	3.4 (0.8); 4	3.6 (1.0); 4
OS_1	First	4.0 (0.5); 4	4.0 (0.8); 4	4.1 (0.8); 4	3.9 (0.6); 4	4.0 (0.7); 4
	Second	3.3 (0.7); 3	3.8 (0.7); 4	3.7 (1.1); 4	4.1 (0.6); 4	3.7 (0.9); 4
	Third	3.7 (0.7); 4	3.7 (0.7); 4	3.7 (0.7); 4	3.6 (0.6); 4	3.7 (0.6); 4
	Total	3.7 (0.7); 4	3.8 (0.8); 4	3.9 (0.9); 4	3.9 (0.6); 4	3.8 (0.8); 4

(continued)

Table 2 (continued)

Items	Year\HEI	Mean (s.d.); Median				
		EEUM	FURB	IFSC	ISEP	Total
OS_2	First	4.1 (0.6); 4	3.7 (1.0); 4	3.8 (0.9); 4	3.7 (0.7); 4	3.8 (0.9); 4
	Second	3.8 (0.8); 4	3.6 (0.9); 4	3.3 (1.2); 3	4.0 (0.6); 4	3.5 (1.0); 4
	Third	3.6 (0.9); 4	3.3 (0.9); 3	3.6 (1.4); 4	3.6 (0.9); 4	3.5 (0.9); 4
	Total	3.8 (0.8); 4	3.6 (0.9); 4	3.6 (1.0); 4	3.8 (0.6); 4	3.7 (0.9); 4
OS_3	First	4.3 (0.7); 4	4.1 (1.0); 4	4.3 (0.7); 4	3.9 (0.6); 4	4.1 (0.8); 0
	Second	3.8 (0.6); 4	4.0 (0.9); 4	4.0 (0.9); 4	4.2 (0.5); 4	4.0 (0.8); 4
	Third	3.7 (0.7); 4	3.8 (0.7); 4	4.1 (0.9); 4	3.8 (0.6); 4	3.8 (0.7); 4
	Total	3.9 (0.7); 4	4.0 (0.8); 4	4.2 (0.8); 4	3.9 (0.6); 4	4.0 (0.8); 4

50% of the students "agree" with the statements proposed (median equal to 4). However, in TPI_4 (*In general, teachers aim to contextualize the contents in a professional perspective*) for practically all curricular years and HEI, in COF_11 (*The engineering course is well organized*) for the third year, and in OS_1 (*I'm satisfied with the school environment and working conditions*), and OS_3 (*In general I'm satisfied with the course*) in a few cases, a median of 3 was obtained indicating a neutral position for more than 50% of the students.

In Table 2, it can also be seen that the two items with the highest score, in average, were OS_3 (*In general, I'm satisfied with the course*) and STI_6 (*In general, the student–teacher interaction is positive*), and by opposition TIP_4 (*In general, teachers aim to contextualize the contents in a professional perspective*) which presents the lowest score.

Figure 1 illustrates the distribution of students' assessment regarding the eleven items under analysis with the corresponding 95% confidence intervals strengthening the item with the highest mean score (OS_3) and the one with the lowest (TIP_4).

In a certain sense the previous results follow the achievements obtained by other authors, namely [1], stressing the importance of teacher and student interaction in the engineering program student's satisfaction, and [3], where students assume that a more practical component helps them to be better prepared for their future career.

These observed differences in average reveal that there was a statistically significant difference in assessment depending on which item is being considered ($F(10) = 406.7$, $p < 0.001$).

In order to understand the differences observed, data were further analyzed by HEI and curricular year.

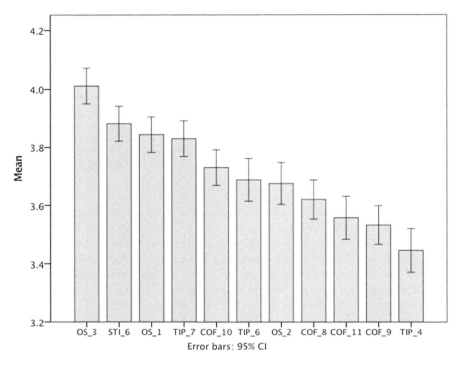

Fig. 1 Distribution of students' answers mean and 95% confidence intervals (CI) for the eleven items (see Sects. 2.2 and 3.1 for acronyms)

5.2 Results by HEI and Curricular Year

For the four HEI, the distribution of students' assessment in the eleven items is shown in Fig. 2.

Some outliers, values considered with an extreme behavior, were observed, namely the ones corresponding to the lower classification 1 "Strongly Disagree" (dots and stars in Fig. 2), in the majority of the eleven items in analysis. Nevertheless, these values were considered for the present study for their importance. In fact, these outliers do not correspond to the same student reflecting a random behavior.

The differences observed, for all the items, among the four HEIs are statistically significant ($p < 0.001$ in all items). So a different behavior is observed according to the item in study. For example, for TIP_4 item, EEUM shows an assessment score lower than the other three HEIs while IFSC has the highest. As another example, in COF_9, item ISEP shows a lower assessment in comparison to EEUM, FURB, and ISFC.

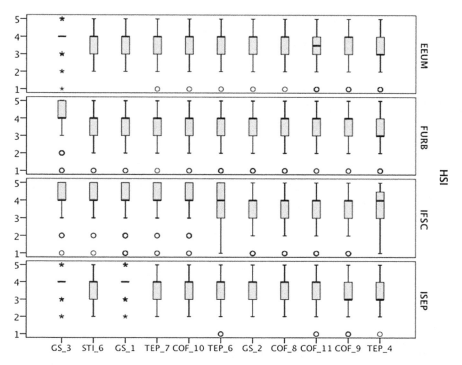

Fig. 2 Distribution of students' answers for the eleven items in analysis by HEI (see Sects. 2.2 and 3.1 for acronyms)

When analyzing the distribution of the scores throughout the three curricular years (Fig. 3), different patterns were observed. Generally, the first year represents the more positive, followed by the second and the third with the lowest score. These differences were statistically significant, except for item COF_9 (*The syllabuses of the courses are well articulated with previously acquired knowledge*) ($H(2) = 2.38$, $p = 0.304$).

Although on different areas of knowledge, other studies also ascertained that students' trust decreases with time [4, 17]. This could justify a higher awareness by teachers of the need and adoption of different and new teaching methodologies reducing student resistance to class activities.

5.3 Relevant Correlations

The student–teacher interaction assessment (STI_6) is positively and significantly related to how students assess positively teachers' performance (TIP_7) ($r_S = 0.60$, $p < 0.001$, Table 3). This result, somehow, might represent that the more students

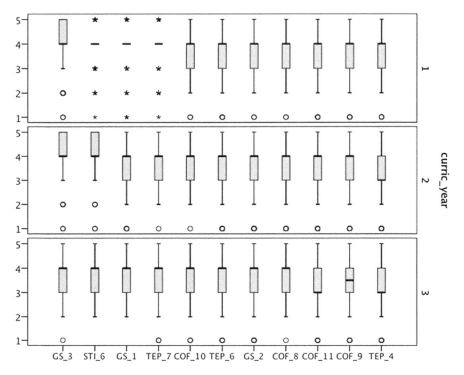

Fig. 3 Distribution of students' answers for the eleven items in analysis by academic year (see Sects. 2.2 and 3.1 for acronyms)

Table 3 Spearman's coefficients, r_s, for the correlations between the items (see Sect. 3.1 for acronyms)

		TIP_7	STI_6	COF_10	OS_1	OS_2
Spearman's coefficient, r_S	TIP_7	1				
	STI_6	0.60*	1			
	COF_10	0.48*	0.46*	1		
	OS_1	0.40*	0.43*	0.43*	1	
	OS_2	0.25*	0.24*	0.31*	0.48*	1
	OS_3	0.50**	0.44*	0.59*	0.60*	0.46*

*Correlation is significant at the 0.01 level (1-tailed)

interact with teachers in class the more positively they assess teachers' performance. This, in some sense, underscores the importance of communication between students and teachers helping the understanding of the students' feedback on teaching [9, 18].

There was a significant positive relationship between the students' overall satisfaction (OS_3) and the students' expectations regarding their course (COF_10) ($r_S = 0.59$, $p < 0.001$, Table 3). This correlation represents a positive result showing that in average students are satisfied with the course and their initial expectations have been met.

5.4   Hypotheses Validation

Table 4 summarizes the results for all hypotheses identifying its veracity.

The students, in average, assessed positively all the items TIP_4, TIP_6, TIP_7, STI_6, COF_8, COF_9, COF_10, COF_11, OS_1, OS_2, and OS_3, allowing to confirm hypotheses 1, 3, 5, 7, 9, 11, and 13 (differences are not significant, see "No" in Table 4). However, this perception changes along the years and between the HEI (differences are significant for hypotheses 2, 4, 6, 8, 10, 12, and 14, see "Yes" in Table 4).

These results show that students' awareness regarding the flow of knowledge changes somewhat over the years and by HEI. At some point, students throughout their course get lost and become unable to interconnect the contents learned.

There is a significant and positive relationship between the students' overall satisfaction and the students' expectations regarding the course. Also, there is a positive relationship between the way students assess their interaction with teacher and how they assess teachers' involvement (differences are significant for hypotheses 15 and 16, see "Yes" in Table 4).

Table 4 Summary of the results regarding the sixteen hypotheses (see Sect. 3.1 for acronyms)

Hypothesis	Item	Significant differences	Hypothesis	Item	Significant differences
1	TIP_4	No	9	COF_8; 9	No
2	TIP_4	Yes	10	COF_8; 9	Yes
3	TIP_6	No	11	COF_10; 11	No
4	TIP_6	Yes	12	COF_10; 11	Yes
5	TIP_7	No	13	OS_1; 2; 3	No
6	TIP_7	Yes	14	OS_1; 2; 3	Yes
7	STI_6	No	15	STI_6; TIP_7	Yes
8	STI_6	Yes	16	OS_3; COF_10	Yes

6 Conclusions

This study presents and discusses some results on students' perceptions in electrical/electronic engineering courses from four higher education institutions: two Portuguese (Instituto Superior de Engenharia do Porto, ISEP, and Escola de Engenharia da Universidade do Minho, EEUM) and two Brazilian (Instituto Federal de Santa Catarina, IFSC, and Universidade Regional de Blumenau, FURB). Six different courses were chosen: two from ISEP, two from IFSC, one from FURB, and one from EEUM. For each of the six electrical/electronic engineering degrees, the first three curricular years were analyzed, corresponding to the first cycle of higher education, with a total of 654 questionnaires considered valid for analysis.

In general students' assessment of all the eleven items considering teachers' involvement perception (TIP), student–teacher interaction (STI), course organization and functioning (COF), and overall satisfaction (OS) is positive (mean higher than 3, in a 5-point Likert scale).

When the results are analyzed by HEI and by curricular year, there are statistically significant differences between the four HEIs and the three years.

Regarding the four HEI, in many items, the best scores are obtained in IFSC. One of the factors that might justify these results is the novelty of the degree (started in 2013), although a more detailed analysis of other possible reasons would be important and necessary.

Along the three years, the students' awareness of their course, in average, becomes more critical, i.e., from one year to the next, the students' assessment, in general decreases slightly. The analysis performed also shows that in general students do not understand the articulation of contents and the flow of knowledge along the curricular years and between curricular units. This might confirm that it is important to include integrating projects in all curricular years, and not only in last years, as it is more usual. This will have an impact on students' teaching/learning process as a first contact with real engineering problems and as a collaborative learning [19–21]. These latter aspects can be an indication that students do not understand the flow of knowledge in the first three years, which strengthens the idea of a further research to be undertaken considering the second cycle of studies (fourth and fifth year).

The students' assessment depends on which item is being analyzed: in general, students are satisfied with the course and with student–teacher interaction (highest score), and by opposition, students perceive that teachers do not contextualize the contents in a professional perspective (lowest score).

The students' overall satisfaction and the students' expectations regarding the course are positively related. Also, there is a positive relationship between the way students assess their interaction with teacher and how they assess teachers' involvement.

The results analyzed in the present work point out that there are no significant differences when considering the two countries, revealing that two countries do not necessarily imply two realities, but instead work with the same objective: to create a space to disseminate, share, and generate knowledge.

Future work will include the analysis of students' feedback in the second cycle of studies in the four institutions. Also, authors are considering the possibility of extending the study to higher education institutions in other countries.

Acknowledgements The authors would like to express their acknowledgments to the higher education institutions and to all the students who accepted, on a voluntary basis, to collaborate in this study.

References

1. Sinclaire JK (2014) An empirical investigation of student satisfaction with college courses. Res High Educ J 22
2. Tasirin SM, Omar MZ, Esa F, Zulkifli MN, Amil Z (2015) Measuring student satisfaction towards engineering postgraduate programme in UKM. J Eng Sci Technol 10:100–109 (Special Issue on UKM Teaching and Learning Congress 2013)
3. Cronje T, Coll RK (2008) Student perceptions of higher education science and engineering learning communities. Res Sci Technol Educ 26(3):295–309. https://doi.org/10.1080/02635140802276587
4. Borges GR, Carvalho MJ, Domingues S, Cordeiro RCS (2016) Student's trust in the university: analysing differences between public and private higher education institutions in Brazil. Int Rev Public Nonprofit Mark 13:119–135. https://doi.org/10.1007/s12208-016-0156-9
5. Lord SM, Layton RA, Ohland MW (2015) Multi-institution study of student demographics and outcomes in electrical and computer engineering in the USA. IEEE Trans Educ 58 (3):141–150. https://doi.org/10.1109/TE.2014.2344622
6. Vaz RÁ, Freira D, Vernazza E, Alves H (2016) Can students' satisfaction indexes be applied the same way in different countries? Int Rev Public Nonprofit Mark 13:101–118. https://doi.org/10.1007/s12208-016-0155-x
7. Alves H, Raposo M (2007) Student satisfaction index in Portuguese public higher education. High Educ Serv Ind J 27(6):795–808
8. Bell JS, Mitchell R (2000) Competency-based versus traditional cohort-based technical education: a comparison of students' perception. J Career Tech Educ 17(1):5–22. https://doi.org/10.21061/jcte.v17i1.589
9. Bagchi U (2010) Delivering student satisfaction in higher education: a QFD approach. In: 7th international conference on service systems and service management (ICSSSM), Tokyo, Japan, 28–30 June 2010
10. Xu H (2011) Students' perception of university education—USA vs. China. Res High Educ J 10. https://doi.org/10.1037/spq0000002
11. Rjaibi N, Rabai LBA, Limam M (2012) Modeling the prediction of student's satisfaction in face to face learning: an empirical investigation. In: International conference on education and e-learning innovations, Tunisia, 1–3 July 2012
12. Carvalho M, Batra B (2015) Pharmacy student's survey: perceptions and expectations of pharmaceutical compounding. Int J Pharm Compd 19(1):18–27
13. Leão CP, Soares F, Guedes A, Sena-Esteves MT, Alves G, Brás-Pereira IM, Hausmann R, Petry CA (2015) Freshman's perceptions in electrical/electronic engineering courses: early findings. In: Proceedings of the 3rd international conference on technological ecosystems for enhancing multiculturality (TEEM 2015), Porto, Portugal, 7–9 October 2015. ACM, pp 361–367. https://doi.org/10.1145/2808580.2808634
14. Leão CP, Soares F, Guedes A, Sena-Esteves MT, Alves G, Brás-Pereira IM, Hausmann R, Petry CA (2016) Sou caloiro de engenharia!: estudo multicaso em engenharia eléctrica/eletrónica/electrotécnica. In: COBENGE 2016, Brasil, 27–30 Sept (in Portuguese)

15. Israel GD (1992) Determining sample size. University of Florida Cooperative Extension Service, Institute of Food and Agriculture Sciences, EDIS, Gainesville
16. Field A (2009) Discovering statistics using SPSS. SAGE, Publications Ltd., London
17. Nguyen KA, Husman J, Borrego M, Shekhar P, Prince M, Demonbrun M, Finelli C, Henderson C, Waters C (2017) Students' expectations, types of instruction, and instructor strategies predicting student response to active learning. Int J Eng Educ 33(1A):2–18
18. Weurlander M, Cronhjort M, Filipsson L (2016) Engineering students' experiences of interactive teaching in calculus. High Educ Res Develop, 1–14. https://doi.org/10.1080/07294360.2016.1238880
19. Alves AC, Moreira F, Lima RM, Sousa RM, Dinis-Carvalho J, Mesquita D, Fernandes S, van Hattum-Janssen N (2012) Project based learning in first year, first semester of industrial engineering and management: some results. In: Proceedings of the ASME 2012 international mechanical engineering congress & exposition (IMECE2012), Houston, Texas, USA, 9–15 Nov
20. de los Ríos I, Cazorlaa A, Díaz-Puentea JM, Yagüe JL (2010) Project–based learning in engineering higher education: two decades of teaching competences in real environments. Procedia—Socd Behav Sci 2(2):1368–1378
21. Stappenbelt B, Rowles C (2009) Project based learning in the 1st year engineering curriculum. In: Proceedings of the 20th Australasian Association for Engineering Education Conference (AAEE 2009), Adelaide, University of Adelaide, Australia, pp 411–416

Celina P. Leão is a Researcher at Centro Algoritmi and has been an Assistant Professor at the School of Engineering of University of Minho since 2003, in Guimarães, Portugal, where she teaches numerical methods and applied statistics. Her main interests are in modeling and simulation of processes and in the application of new methodologies in the learning process of numerical methods and statistics in engineering. Her interests also include engineering gender studies and in mixing qualitative and quantitative approaches.

Filomena Soares received her degree in Chemical Engineering in 1986 at Porto University, Portugal. In 1997, she obtained her Ph.D. in Chemical Engineering at the same University. Since 1992, she works in the Industrial Electronics Department, Minho University, and she develops her research work in R&D Algoritmi Centre. Her main scientific interests are in the areas of system modeling and control, with application to bioprocesses and in biomedical engineering science. Motor and cognitive rehabilitation has been receiving her attention, using serious games and robots to foster the communication with impaired and typically developing children. She is interested in new teaching–learning methodologies, in particular blended learning and virtual and remote laboratories. She supervised several M.Sc. and Ph.D. thesis and was co-author of several scientific articles in international conferences and journals.

Anabela Guedes is a Researcher at CIETI—Instituto Superior de Engenharia do Porto (ISEP), Portugal, and has been a teacher in the Chemical Engineering Department at ISEP since 1995. Her main interests are chemical reaction, process control, and energy optimization. Her interests also include new teaching–learning methodologies.

M. Teresa Sena-Esteves is a Researcher at CIETI and has been a Professor at the Polytechnic of Porto—School of Engineering, since 1993, in Porto, Portugal, where she teaches fluid mechanics and process integration. Her main interests are in chemical engineering, mass transfer, fluid dynamics, separation processes, and swimming pool water treatment.

Gustavo R. Alves graduated in 1991 and obtained an MSc and a Ph.D. in Computers and Electrical Engineering in 1995 and 1999, respectively, from the University of Porto, Portugal. He is with the Polytechnic of Porto—School of Engineering, since 1994. He was involved in several national and international R&D projects and has authored or co-authored +200 conference and journal papers with a referee process. His research interests include engineering education, remote laboratories, and design for debug and test.

Isabel M. Brás Pereira is a Researcher at CIETI—Instituto Superior de Engenharia do Porto (ISEP), Portugal, and has been a teacher in the Chemical Engineering Department at ISEP since 1993, in subjects like heat and mass transfer, instrumentation and control, gaseous emissions treatment. For the last nine years, she has been coordinating the curricular internships of first cycle chemical engineering students. Engineering education is presently one of her main research interests.

Romeu Hausmann is a researcher on Power Electronics in the Department of Electrical and Telecommunication Engineering at University of Blumenau—FURB, Santa Catarina, Brazil. He is a Full-Time Professor at FURB, where he is a lecturer in power electronics and electrical circuits, since 2001. His main interests are in DC–AC multilevel converters, renewable energy, and engineering education.

Clovis António Petry is a researcher at LPPE and has been a lecturer at the Federal Institute of Santa Catarina—Campus Florianópolis (IFSC) since 2006, in Florianópolis, State of Santa Catarina, Brazil, where he teaches power electronics and research methodology. His main research interests are related to teaching electronics and didactical issues, as well as the power electronics research field.

Innovative Methodologies to Teach Materials and Manufacturing Processes in Mechanical Engineering

J. Lino Alves, Teresa P. Duarte and A. T. Marques

Abstract This chapter discusses some methodologies implemented in teaching materials and manufacturing processes at the Department of Mechanical Engineering, of Faculty of Engineering of University of Porto, Portugal, that aim to keep mechanical engineering students motivated and strongly enrolled in classes. Practical classes are structured around experimental works where students have the opportunity to design and perform different experiments, do research using databases, training presentations, do technical reports and posters, and visits to industrial companies. Although the experimental works are very demanding and time consuming, they are extremely appreciated by the students, leading to great motivation for learning and an uncommon enrolment in the curricular units. This chapter presents the methodologies adopted in teaching metallic and non-metallic materials, considering international criteria for engineering students, and learning outcomes and competences. Finally, different cases studies of implementation of this project based learning methodology are presented. These classes contribute to acquire solid technical knowledge and simultaneously, development of soft skills that are extremely important and appreciated by the companies.

Keywords Teaching mechanical engineering · Materials · Manufacturing processes · Innovative methodologies · PBL

J. Lino Alves (✉) · T. P. Duarte · A. T. Marques
University of Porto, Rua Doutor Roberto Frias S/N, 4200-465 Porto, Portugal
e-mail: falves@fe.up.pt

T. P. Duarte
e-mail: tpd@fe.up.pt

A. T. Marques
e-mail: marques@fe.up.pt

© Springer Nature Singapore Pte Ltd. 2018   75
M. M. Nascimento et al. (eds.), *Contributions to Higher Engineering Education*,
https://doi.org/10.1007/978-981-10-8917-6_4

1 Introduction

Teaching with success is a very demanding task, especially in present time, where the students have a continuous contact with the powerful tools of internet and media and nothing seems to surprise them. A rapid knowledge is obtained at the distance of a simple mouse click, independently of the region of the world where the learner is.

Nowadays, when students are admitted to the University, they have already a vast control of specific informatics tools, such as typing texts in word processer and making fancy power point presentations, and many of them are extremely fast at searching for an answer in the World Wide Web. However, when a deeper knowledge about the teaching subjects is required, difficulties start to appear and, essentially when some background about certain scientific principles is demanded, the problems are even bigger [1]. These are the main deficiencies that the authors detected on their students of the Integrated Master Course in Mechanical Engineering (MIEM) from the Department of Mechanical Engineering (DEMec) of the Faculty of Engineering of University of Porto (FEUP)—Portugal, during their classes about materials and manufacturing processes. The introduction of these subjects—materials and technological processes—with the detail that is presently taught is related to the type of mechanical engineers' necessities of the region and the country industrial tissue, and the global employment market. Due to the excellent technical and scientific training that this course (MIEM) provides, proved by the feed-back of national and international employers, in recent years MIEM had a very high demand index, within the panorama of Portuguese public higher education, being enhanced in the last two years, where:

2015—From 995 candidates (representing 2% of 48.306, the total national of public higher education), the number of candidates in 1st option was 374, with a ratio of MIEM candidates/number of slots in FEUP/MIEM equal to 6.2 [2].

2016—From 1059 candidates (representing 2.1% of 49.655), the number of candidates in 1st option was 356, with a ratio of MIEM candidates/number of slots in FEUP/MIEM equal to 6.6 [3].

Due to the increase in the number of students admitted in last years in MIEM/FEUP, and to meet the high demand expectations (Fig. 1) of Portuguese students that want to enter in the public higher education system, the number of experimental classes where students do experiences on their own were reduced.

Considering that and due to the present life style in our *digital society*, students' experimental skills and sensibility for "how to do it by doing" are continuously being reduced, and despite the *WWW* could be a possible milieu for discussion and virtual interaction, the reinforcement of the experimental activities and group work is an urgent need to "materialize" knowledge and experience.

Thus, in order to combat this trend, all the teachers and directors of MIEM, DEMec, and FEUP have been continuously changing its curricula, the teaching methodologies and evaluation system in order to have the courses with an international level, able to train young engineers with appropriate skills to the current demands of national and international companies that want to be globally competitive.

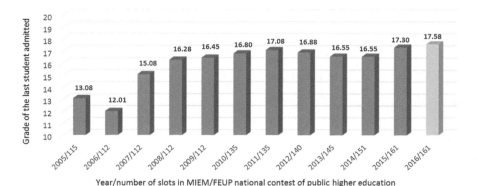

Fig. 1 Evolution along the years of number of slots in MIEM/FEUP and grade (in a scale of 20) of the last student admitted in the course as a result of the national contest for public higher education

In 2006 MIEM was configured according to the Bologna Process, in 2008 obtained the EUR-ACE Accreditation of European Engineering Programs, and in 2013 was ranked 28th in Europe and 92nd in the World in the NTU ranking (National Taiwan University) [4]. In 2016, the MIEM was again accredited by the EUR-ACE, however with some recommendations, being one of them, the increase of experimental classes, resulting from the interviews of the Evaluation Commission with students' leaders [5].

Besides that, FEUP/DEMec and MIEM participate actively in the International exchange programs for both IN and OUT students (Erasmus+, Mobile (Brazil, Latin America and Timor), Protocol with University of Maryland—Baltimore County (EUA), Program Almeida Garrett (Portugal), Projects Mobility Erasmus+ICM: International Credit Mobility, Mobile+2 Merging Voices, Net Magalhães— Program Smile) [6], and have a variety of collaborations with companies, research institutes and different schools worldwide. These exchanges and collaboration programs are an excellent way of implementing new methodologies in teaching, since by receiving students and teachers from other universities one can benefit from their experiences. Collaborations with companies allow the dissemination of the type of skills acquired by students, not only at a technical and scientific level, but also the soft skills that are developed in experimental classes with group works.

One of the main concerns of the mechanical engineering FEUP professors is the transmission of the best general competences CDIO (Conceive-Design-Implement-Operate) [7] and scientific knowledge, and also the development of skills that will be very helpful in the future active professional life.

According to Indicators and Standards that regulate higher education in Europe, the contents of the MIEM (in general), teaching and learning challenges and interaction and collaboration among teachers and students should contribute to the following CDIO skills number [7]:

- 1.2 nuclear knowledge in engineering;
- 1.3 advanced knowledge in engineering;
- 2.1 thinking and resolution of engineering problems;
- 2.2 experimentation and knowledge discovering;
- 2.4 personal skills and attitudes;
- 3.1 group work;
- 3.2 communication;
- 4.4 project.

Presently, the new evaluation system of engineering courses is also based on EUR-ACE skills [8] which for this course are (summary):

- 3.1 Knowledge and understanding—An in-depth knowledge and understanding of the principles of their branch of engineering;
- 3.2 Engineering analysis—The ability to solve problems that are unfamiliar, incompletely defined, and have competing specifications;
- 3.3 Engineering design—An ability to use their engineering judgment to work with complexity, technical uncertainty and incomplete information;
- 3.4 Investigations—The ability to identify, locate and obtain required data;
- 3.6 Transferable skills—Work and communicate effectively in national and international contexts.

The innovative teaching methodologies presented in this work are based on Project-Based Learning—PBL [9–11] and, in addition to the contents described in further sections, this type of evaluation also started to be implemented in 2010/11 in FEUP in the Specialization Course in Design and Product Development and currently in the Master Program of Product and Industrial Design [12–15].

Several researchers and teachers worldwide have also published their experiences using PBL: Panthalookaran and Binu [16], in Rajagiri School of Engineering and Technology, India, also tried something similar to nurture general management skills in their engineering students. Kostal, Mudrikova and Caganova [17], in Slovak University of Technology, Slovak Republic, improved their teaching methodologies through virtual laboratories, enforcing students' capacities to learn by their self-activity and self-responsibility and improving their communication skills. Peréz, García and López [18] in Polytechnic University of Madrid promoted the PBL in their Mechanical/Industrial Engineering courses. Frank, Lavy and Elata [9], in Technion, Israel, implemented the PBL through mini-projects that require the design and construction of devices that perform pre-defined tasks. Zhou [19] used the same type of work to teach manufacturing processes.

In Shamoon College of Engineering, Department of Mechanical Engineering, Beer-Sheva, Israel, Professor Iko Avital, streamlines a competition among students, to design and manufacture a small boat prototype to deliver food and drinks to the tourists in Dead Sea (Fig. 2). Student's teams and teaching staff participate enthusiastically in this project, and seek collaboration and advice from colleagues, professors and technicians.

Fig. 2 Annual dead sea competition of Shamoon College of Engineering, Israel

Meanwhile, some of our DEMec colleagues have also been doing great efforts to introduce these types of methodologies for a more effective knowledge transmission, which encourages extra efforts to keep improving and innovating in teaching methodologies in our materials and technological processes courses.

All these different PBL methodologies are focused on a higher students' responsibility and have a more experimental character with projects to produce or operate specific devices. Although we also have experimental work, a large emphasis is placed on searching scientific data and capacities to clear present ideas and participate in debates, soft skills fundamental to the professional success of young mechanical engineers.

This preoccupation with experimental work, is referred in many recent papers and largely discussed in conferences such as the annual International Conference EDUCON —Collaborative Learning, New Pedagogic Approaches in Engineering Education, organized by the IEEE (Institute of Electrical and Electronic Engineers) [20].

Considering the facts described in this introduction, two examples of classes about materials and processes will be presented, and the innovative methodologies adopted described in detail as well as the final main results achieved that contributed to the improvement of students skills.

2 Evolution of Teaching Methodologies in Materials Classes of MIEM

In the past (before 2006), teaching of Non Metallic and Metallic Materials in MIEM was performed in the classic way with theoretical and practical classes. In the theoretical classes of 1 h + 1 h per week (semester of 12 weeks) for around 130 students at the same time, being impossible to have a personalized knowledge about each student, the emphasis was on presenting the main subjects of the curricular units (CUs). In the practical classes (2 h for 25 students, maximum) students had the opportunity to carry on simple experiments. The final grade was composed by 20% for the reports of practical classes and the remaining 80% for the final exam.

In 2006, when the course was approved by Bologna Process, the teachers of Non Metallic Materials decided to accept the challenge of introducing in MIEM a curricular unit (CU) with the evaluation based on experimental works (PBL), discarding the exams. The classes were changed to just practical ones (with 4 works for ceramics and 3 for polymers) with the goal of giving the student a more responsible and proactive attitude, which is characterized by spending much more time at University/ home studying the main topics taught in classes. Although the contents of the course remain the same, at the beginning of some classes each subject is briefly presented during 15–20 min maximum. After that, students have to answer the questions of the practical works using class facilities and complementary work done at University/ home (the course has 6 ECTS—European Credit Transfer and Accumulation System [21] which corresponds to a total of 162 h (1 ECTS corresponds to 27 h work) of work during the semester, including classes, study, experimental and team work). This methodology was abandoned at the end of two academic years because of the large amount of time needed for teachers to properly assess students, in addition to the time spent in preparing and teaching classes and also requiring too much work time for students to the expected time for this CU.

After 2008, 2 reports for each part: ceramics and polymers, and answers to some handouts to be solved in class or at home are the only responsible items for the final grade obtained. This means that a deeper knowledge has to be obtained about the students from the discussions in all practical classes and continuous contact with the teachers, to obtain a more accurate and fair assessment.

After few years of using these methodologies in teaching and after analyzing the results obtained and comments from the students, the authors introduced in 2010/11 three innovations in order to address some detected deficiencies:

1. One class (2 h) in information literacy in FEUP library about how to use bibliographic databases. This session includes competences in searching in scientific databases (Compendex, Inspect, etc.), integral text, e-books, patents, dissertation and thesis and how to use the Endnote. This specific competences proved to be very helpful for all students, shortening their searching time and obtaining more valuable information and using it on the reports;
2. Seminar of electronic microscopy given by the Materials Centre of University of Porto [22], in order to give the students specific tools for microstructure analysis of all types of materials;
3. Introduction and use of CES Edupack software from GRANTA [23]. This software is very important to search information about materials and manufacturing processes and to relate properties with the specific production processes for all materials.

In relation to the CU of Metallic Materials, there were no changes since the implementation of the Bologna process in relation to the type of classes, and the theoretical classes were held for 1 h + 1 h per week and the practical classes of 2 h per week for groups of about 22–24 students. This CU has distributed evaluation (experimental work, preparation of written report, poster, presentation and oral defense—30%) with final exam (70%). In this CU, the innovation, described later, is to propose to the students more challenging experimental works, identification of metallic components, instead of experimental works about a specific heat treatment.

In the work presented here it is not intended to discuss the contents of the CUs, since in all the external evaluations to which MIEM has been submitted, the evaluation panels refer the fact that the subjects taught are adequate to the skills that a mechanical engineer should have.

This chapter presents innovative teaching methods where students are expected to acquire the contents of CUs by searching in high quality bibliographic databases, consulting technical books, viewing videos, or any other sources of information available nowadays.

The following sections describe the objectives of the assignments, all tasks performed by the students, and the evaluation processes.

3 Methodology

The experiments reported were developed in CUs related to the various materials available to the mechanical engineer more, in particular, metallic and non-metallic materials. It is intended that at the end of these CUs (together with a general CU of materials science taught in MIEM's first year) students acquire solid knowledges in this area, that can be used in CUs of manufacturing processes, design and others, and also to provide them with tools to solve all the challenges they will encounter in these subjects during their professional life.

3.1 Non Metallic Materials

Learning outcomes and competences

At the end of the semester (3rd year, 1st semester), the students should have acquired basic and advanced engineering knowledge about ceramics, polymers and polymers matrix composites, namely:

- Knowledge about the different ceramics, polymers and composites, used in the different brands of engineering, main applications and properties;
- Be able to understand the mechanical, optical, thermal and electrical properties of these materials;
- Be able to select the most suitable materials considering the desired application;
- Capacity to perform different types of experimental work to collect data, interpretation and relations with the learned subjects;
- Perform small experimental projects using the learned materials, namely materials selection and production processes;
- Capacities to collect and organize scientific information, using books, scientific papers, internet, databases, technical visits or interviews, elaboration of technical reports, posters and public oral discussions and presentations;

- Capacity to do practical team works, presentation and discussion of the results obtained [24].

In order to achieve the learning outcomes and competences, different experimental works have been proposed. In this work only two examples are presented, one in the area of ceramic materials and another in the area of polymeric and composite materials. Other experimental works already used in this CU are described in other publications [11, 25].

3.1.1 Ceramic Materials

Analysis and interpretation of a scientific paper

Objective:

Analysis, interpretation of a given scientific paper about ceramic materials and complementary search about the topic developed on the paper and elaboration of a report, a poster (A3 or A4), a presentation and public debate about the performed work.

The main challenge proposed to the students (groups of 2–3) is to do a report that contains the necessary information for the reader to take a decision:

> Consider that you are an employee in a company and that your boss asks you to study a subject and supply him/her with a report containing all the necessary state of art information to take a decision about adopting or modifying a technology/process in the company.

The scientific papers provided by the professors are, generally, all of the same year of publication, coming from reference journals about ceramic materials, with the same degree of difficulty of interpretation, identical number of pages and related to the CU contents. A guide, presented below, is also given to students, with detailed information on the preparation of this work.

Guide:

The reports should be quickly understood by the reader. Therefore they should:

- Be well presented (the subjects being well organized enhancing what is more important);
- Be well written and not contain spelling mistakes;
- Present the subjects obeying a scheme defined at the beginning. Thus, after the cover sheet, they should include the index showing the organization of the report;
- Use frequently graphics, tables, figures or others that turns the presentation appealing, easy to read and to understand the work performed;
- Indicate the main conclusions at the end;
- Identify the references, by names and dates, or numbers, on the text, figures, tables and graphics;
- Use SI Units.

The presentation of samples or parts/components of the studied materials, during the oral presentation, as well as personal initiatives to visit companies or interviews to specialists, related with the proposed topic will be graded positively.

Evaluation:

All the groups should deliver the report till the deadline (indicated at the beginning of semester) and supply on the 1st day of the oral presentations and debate a file containing the following elements:

- Presentation of the work;
- Poster;
- Report;
- Elements collected during visits or others.

Not obeying the deadline to deliver all the work elements will be negatively classified. All the reports presented by the students that contain parts from other reports will be graded with "0".

The single use of internet sites as references will be classified very negatively. All the groups have to present in annex at least the three best scientific papers (copies) found about the studied topic (Warning: these papers should be used as references on the report). Do not forget that there are in the FEUP library, Databases, such as Compendex and the knowledge library: http://www.b-on.pt, where numerous papers can be found.

Detailed instructions about reports structure:

The basic structure of the reports (to be adapted for each particular paper) should be the following:

- Cover sheet: Authors of the report (complete names), local, period of the work and due date, subject and course, work title, number of the group and class, reference to supervisors and main collaborators.
- Contents: Include page numbers and all the detailed titles indicated along the work.
- Summary and Objectives: The objectives and working methods employed should be clearly indicated.
- State of Art: Comprehension and discussion of the following aspects (adapted according to each paper subject and relations with the contents of the curricular unit):

  - Typical chemical composition, type of chemical bonds, structure, etc.;
  - Powder manufacturing processes;
  - Physical and mechanical properties, or others;
  - Processing (manufacturing processes for parts and components);
  - Applications (practical examples in different areas);
  - Future and new challenges;
  - Other elements that seem interesting (for instance, recycling possibilities).

- Conclusions: Present the main conclusions in a clear synthetic way.
- Future work suggested and criticisms: when justified, the difficulties found and suggestions concerning performing future work, working methods, topics, etc., should be indicated.
- References: The incorrect indication of the references penalizes significantly the work. Each reference or paper should be indicated in brackets along the text, using the last name of the first author and publication date, or alternatively by a number. At the end of the work, each author cited will have the complete specification of the reference, including:

Author(s), title, editor (or journal where the article is included), data, local of edition and pages

In case the reference was done by two or more authors the abbreviation et al. can be used in the text, but at the end all the authors have to be referred.

Example:

Reference during the text:

(Duarte et al. 2008) or [1]

Reference in the bibliographic references:

[Duarte et al. 2008] Teresa P. Duarte, Rui J. Neto, Rui Félix, F. Jorge Lino, "Optimization of Ceramic Shells for Contact with Reactive Alloys", Trans Tech Publications, pp. 157–161 (2008);

   or

Teresa P. Duarte, Rui J. Neto, Rui Félix, F. Jorge Lino, "Optimization of Ceramic Shells for Contact with Reactive Alloys", Trans Tech Publications, pp. 157–161 (2008) [1].

Presentation and Oral Debate:

The schedule of oral presentation of the work is defined at the beginning of the semester. The maximum time for the presentation is 8 min for each group (exceeding this time has a penalty) followed by a debate (around one hour) with all the students that did the same work (see Fig. 3).

Fig. 3 Schematic of the debate

The evaluation of each group element is based on the following:

- Time used during the presentation;
- Presentation structure;
- Knowledge of the subject, capacity of making a presentation and answering questions about it.

The questions of other group colleagues and the teaching staff are helpful to enhance the debate. The performance of each student will be evaluated by the teachers of the curricular unit and by the students.

Poster:

The poster is evaluated considering the inclusion of the following elements:

Design; Subject title; Course; Year; Objectives; Introduction; Work done; Conclusions; Future work; Photo of the groups elements; Place of the work; and other elements considered relevant.

In order to encourage students to produce high quality posters both in terms of content and design, in the last 4 academic years (since 2013/2014), teachers have decided to launch a competition and to award a diploma the 3 best posters each year which are displayed in the DEMec's standpoint (Fig. 4) in the following two academic years, and serve as an example to the students of MIEM.

Figure 5 shows two posters of this subject; poster (a) is considered a good one, while poster (b) had a lower grade (it does not have the period of the work, course, objectives and conclusions, does not explain the topic of the paper, no captions and has a poor design).

Grade:

The final grade is obtained by the evaluation of three main items:

1. Report (11/20):
   Cover sheet (1/11); Contents (0.5/11); Summary and objectives (0.5/11); State of Art (4/11); Conclusions (1/11); Future, criticisms and annex (0.5/11); References (1/11); The three best scientific papers and their use on the report (2/11); Design of the report (0.5/11);

Fig. 4 Posters' displayed in DEMec aisle and classroom

(a)   (b)

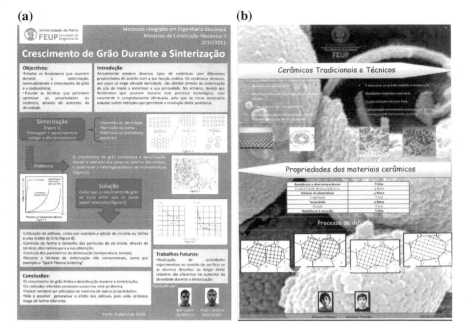

Fig. 5 Posters of the work **a** poster with a good grade and **b** poster with lower final grade

2. Poster (3/20)
3. Oral presentation and debate (6/20):
   Oral presentation (3/20); Debate (3/20).

In the following, the papers from International Journal of Applied Ceramic Technology provided to students in the 2016/2017 academic year are presented, where one can see that the covered topics to be studied are actual and pertinent.

1. Robert Gmeiner, Gerald Mitteramskogler, and Jurgen Stampfl, Aldo R. Boccaccini, "Stereolithographic Ceramic Manufacturing of High Strength", Int. J. Appl. Ceram. Technol., 12 [1] 38–45 (2015).
2. Uwe Scheithauer, Eric Schwarzer, Hans-Jurgen Richter, and Tassilo Moritz, "Thermoplastic 3D Printing—An Additive Manufacturing Method for Producing Dense Ceramics", Int. J. Appl. Ceram. Technol., 12 [1] 26–31 (2015).
3. Jie Yin, Zhaoquan Zhang, Zhengren Huang, Hui Zhang, Yongjie Yan, Xuejian Liu, Yan Liu, and Dongliang Jiang, "Aqueous Gelcasting and Pressureless Sintering of Zirconium Diboride Ceramics", Int. J. Appl. Ceram. Technol., 11 [6] 1039–1044 (2014).
4. Guosheng Xu, Tomohiko Yamakami, Tomohiro Yamaguchi, Morinobu Endo, and Seiichi Taruta, Isao Kubo, "Pressureless Sintering of Carbon Nanofibre/SiC Composites and Their Properties", Int. J. Appl. Ceram. Technol., 11 [2] 280–288 (2014).

5. Michael C. Tucker, Jay Tu, "Ceramic Coatings and Glass Additives for Improved SiC Based Filters for Molten Iron Filtration", Int. J. Appl. Ceram. Technol., 11 [1] 118–124 (2014).
6. Amnon Rothman, Sergey Kalabukhov, Nataliya Sverdlov, Moshe P. Dariel, and Nahum Frage, "The Effect of Grain Size on the Mechanical and Optical Properties of Spark Plasma Sintering-Processed Magnesium Aluminate Spinel $MgAl_2O_4$" Int. J. Appl. Ceram. Technol., 11 [1] 146–153 (2014).
7. Ramanathan Papitha, Madireddy Buchi Suresh, and Roy Johnson, Das Dibakar, "High-Temperature Flexural Strength and Thermal Stability of Near Zero Expanding doped Aluminum Titanate Ceramics for Diesel Particulate Filters Applications", Int. J. Appl. Ceram. Technol., 11 [4] 773–782 (2014).
8. Preeti Bajpai and Parag Bhargava, "Effect of Heat Treatment Schedules and Glass Powder Particle Size on Glass Infiltration in Porous Alumina Preforms", Int. J. Appl. Ceram. Technol., 11 [3] 543–549 (2014).

The evaluation of this work has not always been the same in all the academic years, having varied according to the total number of students, with the analysis of the results obtained in each year and also with improvements proposed by the students.

In the last three academic years, the debate and oral presentations were no longer held due to the high number of students enrolled in this CU as a result of the increase in the number of students joining the MIEM (Fig. 1). The evaluation of this work was carried out only by the delivery of a written report, analyzed with much more rigor and giving a great deal of relevance and quotation to the capacity of synthesis, ability to present other scientific works carried out in the same area of knowledge and great exigency in the quality and presentation of the bibliographic references used to make the report.

The ability to perform oral presentations, debates and posters is evaluated in another experimental work proposed in the part of ceramic materials with the title: Production of ceramic components.

Results Analysis

Reports:

The analysis of the reports delivered by the students has shown that they can produce a very well structured report, with very high graphical quality (cover sheet, figures and tables, typing font and layout of the pages). This means that in general the reports are pleasant to read and the main conclusions and important data are very easily and quickly obtained.

The weak points detected in the reports are:

• Many students still have difficulties in indicating the sources of the data, figures and tables used on the report, although very precise instructions were supplied to them (this has been improving over the years);

- Some did not understand the correct way to indicate the references along the text although they introduce the references at the end in a correct and complete way;
- There is still a tendency to use as references, a considerable number of websites. Although this is not bad, because a lot of useful information can be obtained, it is not enough for engineering students;
- The great majority selected the three papers that they considered the most important and included them in the Annex, but they did not use the concepts/ ideas contained in the papers in the report. This means that this capacity to extract the most important data (synthesis capacity) from a subject that is studied and explained in detail is still a lack in students' capacities.
- Some of them complain about the difficulties in understanding technical English;
- Most of the students focus some innovative tendencies for the future in the subject presented in the paper, but many of them forget to check if the authors have published any other papers after the current one. This is a very important issue, considering that in some years, not all the papers submitted to the groups were from the same year, and many innovations could occur after the supplied paper.

Posters:

In general, students design posters in accordance with the supplied instructions. Some had a lower grade because they do not have the period of the work, course, objectives and conclusions, do not explain the topic of the paper, do not contain captions and has a poor design (Fig. 5).

Oral presentation and debate:

- Students are still not very comfortable with this type of evaluation, and many of them tend to almost not raise questions to the colleagues, because they fill inhibited and are afraid of what the colleagues can think about them;
- Some students are very active and participate intensively, but many times they just talk about generalities, when the teaching team ask about more detailed aspects of the work, and specially topics where it is necessary to relate the things that they read with the contents of the CU, they have serious difficulties;
- Although students have to evaluate other students' presentations and discussions, they tend to give very high grades to all of them and not distinguish the ones that really know the contents of the CU. However, this has been corrected recently with the amount of points that the students can distribute in their grading.

From all the work performed by the students, one can summarize the following points:

- Some difficulties still persist to transmit to the students groups the rigor of the assignment and their responsibility in creating the necessary conditions to independently conduct the work to reach the course goals.
- After some years of implementation of this type of work, we figure out that students are improving and start to be more familiar with this type of challenges. This is the only class, during their Integrated Master's Course in Mechanical Engineering, where they are confronted with this type of continuous evaluation.

We asked some students to give their opinion about this practical work, and the main points can be summarized as:

- Difficulties in reading and understanding technical English, but in the end they considered that they reached significant improvement. This is the main difficulty and we are strongly convinced that this is the reason why they do not include more scientific papers information on the reports;
- Short period of time to perform the work, considering the requests that they had at the same time for other CU;
- Difficulties in collecting information about more technical aspects, due to not finding the correct papers and also because some of the papers that they considered interesting, based on the available abstracts, were not of free access and signed by the school library;
- Not many books available about ceramics;
- They liked the challenge for the deep study of the ceramic topics, considered the supplied papers interesting, learned a lot and should even have more time to better study the subject;
- The work contributed to their synthesis capacity, and a systematic way to study a subject.
- An excellent challenge placed in the curricular plan in the middle of the course and that allows to prepare the way in which a master's thesis must be elaborated.

After 9 years of schooling to challenge the 3rd year students of the MIEM with this work, some changes have been made, some by students' suggestions, others because now there are more students than at the beginning of this pedagogical experience (Fig. 1). As an example, it was decided in the next academic year (2017/18) to change some items of the guide given to the students, namely: definition of the maximum number of pages that the reports must have (development of synthesis capacity), request the presentation of the summary of the scientific paper delivered to each group in the form of a schematic, with a maximum of two pages and elaboration of a mind map with the main topics of the article and that indicates the main questions that it raises, in order to perceive more easily the strategy followed by each group.

3.1.2 Polymers and Composites

Polymers and composite materials of polymeric matrix: from base science to engineering applications

Objective:

Perform a bibliographic study (monograph) of the subject aiming the learning of polymers and composite materials of polymer matrix, from the chemical part, to their mechanical behavior, transformation processes and applications.

To accomplish the above, students (in groups of 2–3) should do research on technical-scientific articles and bring them to the discussion in class. This should be done through oral presentations in class, report and final presentation (5 min) followed by a final discussion. In the end of semester, they should elaborate a monograph on the subject, taking into account all the suggestions and indications given by teachers and colleagues during preliminary presentations.

Students should follow the same instructions given in the work on ceramic materials (Analysis and discussion of a scientific paper) as regards the preparation of reports and correct presentation of bibliographic references. It is intended that every year the subjects were different, up-to-date and suggestions from students who demonstrate particular interest by a certain subject that fits the contents of the CU, are always accepted.

As an example, some of the themes given in 2016/2017 are indicated below:

1. Methodology for design of moulds in polymeric matrix composites
2. Methodology for design of dies in polymeric matrix composites
3. Characterization of polymers and polymeric matrix composites for bicycle wheels/tyres
4. Characterization of polymers and polymeric matrix composites for car wheels/tyres
5. Processing of polymers and polymeric matrix composites for Formula 1 helmets
6. Polymers and polymeric matrix composites for sustainable development: waste-for-life
7. Design of musical instruments in polymers and polymeric matrix composites
8. Selection of polymers and polymeric matrix composites for pneumatic circuits
9. Selection of polymers and polymeric matrix composites for hydraulic circuits
10. Methodology of short and long-term design of pressure vessels made with polymers and polymeric matrix composites
11. Water assisted injection moulding: process simulation
12. Injection moulding of polymeric matrix composites: process simulation
13. Thermoforming/"stamping" of polymers and polymeric matrix composites: process simulation
14. Extrusion of polymers and polymeric matrix composites: process simulation

15. Hot plate press of polymers and polymeric matrix composites: process simulation
16. Polymers and polymeric matrix composites for energy generation (including polymeric trees and plants)
17. Selection of polymers and polymeric matrix composites for aerospace and aeronautic industries
18. Polymers and polymeric matrix composites for high temperatures: raw materials, processing and characterization
19. Polymers and polymeric matrix composites for adaptive structures: applications, characterization, processing
20. Recycling of polymers: recycle, reprocess, reuse
21. Recycling of polymeric matrix composites: recycle, reprocess, reuse
22. Processing of polymers and polymeric matrix composites for electric and electronic industries
23. Conductive polymers and polymeric matrix composites: how to enhance electrical and thermal conductivity
24. Manufacture of vessels/tanks in polymers and polymeric matrix composites: process simulation
25. Natural polymers and polymeric matrix composites: origin, processing, characterization, end of life
26. Biomimetic applied to polymer development
27. Biomimetic applied to the development of polymeric matrix composites
28. Challenges for polymers and polymeric matrix composites in the shoe industry
29. Project specificities with polymers and polymeric matrix composites
30. Selection of polymers and polymeric matrix composites for offshore wind energy
31. Polymers and polymeric matrix composites for "Additive manufacturing": raw materials, processing, applications
32. Polymers and polymeric matrix composites for 3D "printing": raw materials, processing, applications
33. Cold press of polymers and polymeric matrix composites: process simulation
34. Prediction of long-term behaviour of polymers and polymeric matrix composites
35. Fatigue behaviour of polymers and polymeric matrix composites
36. Creep and stress relaxation of polymers and polymeric matrix composites
37. Stress corrosion of polymers
38. Toughness of polymers and polymeric matrix composites
39. Life cycle analysis of polymers and polymeric matrix composites
40. Permeability of gases in polymers and polymeric matrix composites
41. Influence of humidity in the short and long-term mechanical behaviour of polymers and polymeric matrix composites
42. Influence of temperature in the short and long-term mechanical behaviour of polymers and polymeric matrix composites

43. Influence of aggressive liquids in the short and long-term mechanical behaviour of polymers and polymeric matrix composites
44. Polymers and polymeric matrix composites for low velocity impact
45. Methodology of design of gears in polymers and polymeric matrix composites
46. Methodology of design of bearings in polymers and polymeric matrix composites
47. Processing of elastomers
48. Methodology of design with elastomers
49. Strain rate sensitivity of polymers and polymeric matrix composites
50. UV sensitivity of polymers and polymeric matrix composites
51. Joining processes of polymers and polymeric matrix composites
52. Deployable structures with polymers and polymeric matrix composites
53. Simultaneous influence of temperature, humidity and aggressive environment in short and long-term behaviour of polymers and polymeric matrix composites
54. Short and long-term biocompatibility of polymers and polymeric matrix composites: characterization, methodology
55. Smart polymers and polymer matrix composites: types, characterization, processing
56. Polymers and polymer matrix composites for high speed impact
57. Wear behaviour of polymers and polymeric matrix composites
58. Machining of polymers and polymeric matrix composites
59. Polymers and polymeric matrix composites with low friction coefficient: applications, characterization, processing
60. Blow moulding: process simulation
61. Processing of polymeric matrix composites
62. Polymers and polymeric matrix composites with functional gradient
63. Polymer alloys
64. Hybridization polymeric matrix composites
65. Characterization of polymers and polymer matrix composites for tires "not" tyres (NPT-non-pneumatic tyres)
66. Multi material injection moulding
67. Manufacturing tolerances of polymers and polymeric matrix composites
68. Gas-assisted Injection Moulding: process simulation
69. Welding processes of polymers and polymeric matrix composites
70. Design of mechanical polymer connections and polymeric matrix composites
71. Design of a "skate" in polymers and polymeric matrix composites
72. Design of a shelf in polymers and polymeric matrix composites
73. Design of a Coca-Cola bottle
74. Design of a plastic bag
75. Fire behaviour of polymers and polymeric matrix composites
76. Polymeric foams: applications, characterization, processing
77. Design of hyperplastic polymers and polymeric matrix composites
78. Cleaning tools used with polymers and polymeric matrix composites
79. Electromagnetic properties of polymers and polymeric matrix composites

Results Analysis

Regarding the reports submitted by the students from 2008 till 2017, the deficiencies and difficulties experienced allow us to draw the same conclusions already presented regarding the reports on ceramic materials—analysis and discussion of a scientific article.

Classroom presentations of the theoretical work have been carried out without difficulty and most of the students present a very good level of quality. These presentations are also a preparatory work for the master's thesis defenses and presentations that are often necessary in a work context.

The monograph proved to be a very useful tool for both students and teachers. The formers allowed the consolidation of the concepts discussed and led to the creation of new links between them, since the same issue is approached from different perspectives. For professors allowed the formative evaluation and the detection of difficulties in understanding concepts and the opportunity to again explain them, individually or in large groups. The writing of the monograph is seen as a relevant and effective preparatory work for more in-depth monographs, such as the master's dissertations to be presented and defended at the last semester of the MIEM.

The main problem identified during the last years was the late start of the writing of the monograph which is reflected in some lack of organization and verification of the contents. The main reason may be the inertia to begin the writing exercise, that only disappears gradually with continued practice and, still the difficulty that some students feel in organizing and summarizing high amounts of information.

3.2   Metallic Materials

Learning outcomes and competences

It is expected that in the end of CU students will be capable of understanding and anticipate steel and cast irons microstructures based upon chemical composition and heat treatments. Also they must be able to relate microstructures with mechanical properties such as strength, ductility and toughness. It is expected that students know and understand main delivery states of metallic alloys and the meaning of their heat treatments. Finally, they must be able to do materials selection based upon their mechanical and technological properties. Also they must be able to choose or specify heat treatments based on predefined objectives [26].

The CU of Metallic Materials (MM) of MIEM, second year, second semester, has 6 ECTS and a total of 56 h of contact (28 h theoretical and 28 h practical, both of 2 h classes per week). The course demands a total 162 h work, divided among classes, exams, study, development of the practical work outside the classes, and a technical visit to a company that is a steel heat treater and a commercial consultancy and steel distributor (Ramada, Ovar, Portugal) which cooperates with the CU, supplying leftovers of steels for the metallographic samples and steels to make the samples for the mechanical tests.

Experimental work:

The practical classes are taught in a materialographic laboratory by different professors with the support of a technician, for maximum of 22 students in each class (around 160 students in total). The first 4 classes are intended to present the subject covered on the experimental work. The idea of the classes is to supply the necessary tools for students' startup with the work. During these classes some handouts are given to students as well as some exercises and assignments to solve during the classes.

The remaining practical classes are entirely dedicated to the experimental work, done in groups of two students (exceptionally 3). Although they may choose topics outside the steels world, the normal procedure is working with this group of materials because the heat treatments are more complex and just to use the metallographic consumables for steels during samples preparation for metallographic analysis (no contamination if someone is using a softer material).

The typical eight topics covered are:

1. Quenching and tempering
2. Austempering
3. Normalizing
4. Spherodization
5. Temperability determination—Jominy test
6. Carburizing
7. Charpy test in normalizing and tempered condition
8. Identification of real components made of ferrous alloys

Students receive one handout with all the instructions to elaborate the final report and the A4 poster, the due date and the final presentation and discussion dates. The document also includes a set of instructions of how to write a technical report. These instructions cover the following items:

- Title of the work;
- Contents;
- Abstract and objectives;
- Introduction;
- Literature review;
- Study of the supplied material (prediction of the hardness and microstructures and experimental analysis of these two parameters and discussion);
- Study of the heat treatment (definition of the heat treatment cycle, forecast of the hardness and microstructure, experimental determination of these two parameters and discussion);
- Conclusions;
- Future work;
- References;
- Annexes.

The other supplied document is specific for each group, according to the type of the experimental work selected from the 8 topics above referred. This document focuses on the following items:

Objectives and general methodologies:

- Identification of the steel group;
- Characterization of the group through typical properties, applications, manufacturing processes and other important elements;
- Individualization of the material inside it's group. Comparison with other steels of the same group;
- Study of the possible microstructure in the as supplied state. Relation with the forecast metallographic state:
- Draw a schematic of the forecasted microstructure;
- Confirmation of the schematic through the samples already prepared;
- Program the experimental work to be developed. Define the heat treatments and anticipate the final microstructures and hardness;
- According to the heat treatment cycles defined, do the adequate treatment to all the samples;
- Confirm the results obtained by comparing them with the predicted ones;
- Design and perform adequate complementary tests to clarify some doubts relatively to the results obtained.

The type of experimental work, as specified before, can cover different heat treatment cycles, but in generality, each group receives 3 samples (exception for Jominy and Charpy tests). One of the samples is kept in the "as supplied state", and the other two are intended to study the effect of one heat treatment parameter, for example temperature, dwell time at the heat treatment temperature, cooling medium, protective atmosphere, or other interesting factor.

The first seven types of work have clear rules and indications, and students are more or less conducted to the final result. On the other hand, the work number 8 (Identification of real components made of ferrous alloys) is freer, considering that students have to bring to the classes one component that they are curious about the material that was used in its construction [27]. This type of work is more demanding, and is usually chosen by the students that have a more practical intuition or are more adventurers to accept this challenge. Many times they can choose components that are available at classes and that are the result of visits to companies or projects with the industry. The goal of this work is using all the acquired knowledge, professors experience and all the available experimental facilities in DEMec FEUP, to define and execute several heat treatments that can give inputs to help in understanding and identifying the possible type of steel (metallic material) used in that specific application.

Considering this, one presents some of the details of this last type of work.

The following are some of the material identification works that were done in recent years:

1. Steels used in ancient Portuguese bridges (D. Luis, Pinhão, Trezoi and Viana), see Fig. 6. When these bridges were repaired, the degree of steel degradation was evaluated in FEUP and some leftovers were kept in the materialographic laboratory. Figure 7 shows some fractured tensile specimens machined from the Trezoi Bridge.
2. Steels for files (Fig. 8 left)
3. Circular saws (Fig. 8 right)
4. Racing car transmission shafts (Fig. 9)
5. Gears and brakes (Fig. 10);
6. Knives and blades (Fig. 11)
7. Springs
8. Tools and others.

The work starts with the characterization of the supplied part (measurements, pictures, and search data on books and www) on the "as supplied condition". Cut of samples for analysis. When they are supplied in the "as treated" condition, samples

Fig. 6 D. Luis (Porto) and Pinhão (Pinhão) bridges

Fig. 7 Tensile specimens machined from the Trezoi Bridge

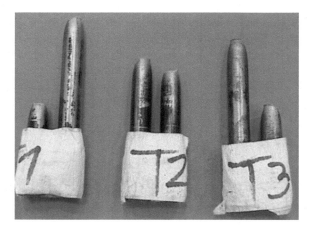

Fig. 8 Files and circular saws

Fig. 9 Racing car transmission shafts axles

Fig. 10 Break system

are cut with an abrasive disc with abundant refrigeration to avoid heating the part and change the hardness.

After this phase, and in case of having small pieces, the samples are cold or hot (preferably, because it is fast) mounted in a thermosetting resin support to allow an easy hand polishing. The polishing sequence adopted is water grinding with SiC paper abrasives (grits #80, 180, 320 and 800), followed by polishing with clothes impregnated with alumina and diamond (3 and 1 μm), respectively.

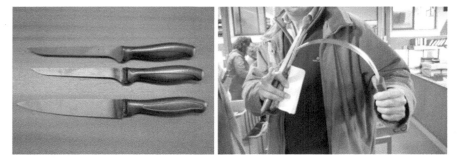

Fig. 11 Kitchen knives in martensitic stainless steels

The metallographic attack is usually done with Nital 2% (2% nitric acid and 98% alcohol). When necessary, a specific reagent is prepared according to the recommendations of Metals Handbook [28] or other books. Samples are then analysed in an optical microscope (Zeiss, Germany) and digital micrographs are obtained. In some cases, if needed, electronic microscopy (usually SEM and microprobe analysis) can be performed at the Centre of Materials of University of Porto [22].

The hardness is determined using Brinell, Rockwell C or Vickers (also micro hardness) scales.

After this point, and using books, as good examples one can cite references [29–35], catalogues (Ramada [36], Thyssen, SSAB and others), CES Edupack Software [23], in house performed dilatometry tests and diagrams of isothermal and continuous cooling transformations, available at the laboratory or other internet data, students' define the heat treatments to perform in the heat treatment laboratory furnaces. The heat treated samples are then polished and etched for microstructural analysis and hardness determination. Figure 12 shows some of the stages during the experimental work.

In the presence of all the results obtained, students compare the experimental results with the theoretical ones, and if any doubt still exists, they can repeat some of the heat treatments to confirm some of the data.

After this stage they start to elaborate the report in the classes, under professors' supervision, and prepare themselves for the final presentation and discussion.

The reports are corrected by the professors that during the final presentation (7 min for each group, followed by an 8 min period of discussion) ask questions about the report but also about the entire subjects taught in the classes. During the discussion other students are also encouraged to participate and are asked questions by their colleagues or teachers.

After the end of classes a simple inquiry, using the Google Drive tool was produced. Students were free two answer, and 25% of them collaborated in this work, with 90% thinking that the experimental works are very important to learn the topics covered on theoretical classes, with 10% saying that is few important. None said that is no important.

Fig. 12 Visit to a steel company, experimental classes and heat treatments laboratory

Students considered the experimental works very challenging, although some of them said that they finished the classes with a deep knowledge about their specific steel but without much knowledge about other steels. We cannot agree with this comment because in the end of semester, students have public presentations of the work and can raise questions and participate actively in this process. In our point of view they are not paying attention to other presentations because they are just worried with their own work.

Students considered that the time available to do the experimental work is adequate and that the identification of components is challenging, but very demanding, needing an even more intense collaboration of professors and laboratory technicians.

4 Conclusions

The use of the presented methodologies contributes significantly to the students of the MIEM to acquire the CDIO and EUR-ACE competences indicated for mechanical engineers.

The introduction of the Project-Based learning assessment in the Non-Metallic Materials Curricular Unit of the Integrated Master's Degree in Mechanical Engineering of the Faculty of Engineering of the University of Porto has successfully changed the way students study and learn the contents related to ceramic, polymeric materials and polymer matrix composites.

It has been shown that evaluation based on a practical work of a scientific paper about ceramic materials or on a current topic about polymer or composite materials that requires considerable research on the World Wide Web and databases of scientific articles, followed by the elaboration of a report, an oral presentation and a public debate is a very demanding job that the students appreciate and actively participate. This structure facilitates their learning and enrollment in the curricular unit, increasing their knowledge about the materials studied and contributes to the development and improvement of their research and synthesis capacities, in writing reports and doing presentations that they will constantly need in the near future.

The introduction of a strong experimental component on the materials classes about metallic materials promotes a great students enrolment on the classes with excellent final practical grades that are responsible for the elevated percentage of final approvals in the MM classes of MIEM's Master program.

Students appreciate the type of experimental work proposed, because they have to do, themselves, samples preparation, microstructural analyses, mechanical tests and heat treatments, interpret, present and discuss the results obtained. They can propose experimental works with materials related to mechanical construction and with the ones that they, from any reason, were involved in their use.

This type of work, although very demanding, is recognized by the students that it is extremely motivating, offering them an excellent opportunity to learn by doing. These classes have been evolving for more than 30 years and when present engineers are asked about the memories of the course, they all seem to remember their materials classes' work they have done. Nothing is more rewarding for a teacher than the perennity of is teachings and influence.

Acknowledgements Authors gratefully acknowledge all the MIEM students that participate in this project, and the funding of Project NORTE-01-0145-FEDER-000022—SciTech—Science and Technology for Competitive and Sustainable Industries, co-financed by Programa Operacional Regional do Norte (NORTE2020), through Fundo Europeu de Desenvolvimento Regional (FEDER).

References

1. Henning K, Bornefeld G, Brall S (2007) Mechanical engineering at RWTH Aachen University: professional curriculum development and teacher training. Europ J Eng Educ 32(4):387–399
2. Admission in FEUP in 2015/16, Internal report, Support Unit to the FEUP Administration, FEUP, Porto, Portugal, November 2015
3. Admission in FEUP in 2016/17, Internal report, Support Unit to the FEUP Administration, FEUP, Porto, Portugal, November 2016

4. NTU Ranking (2013) http://nturanking.lis.ntu.edu.tw/Default.aspx. Accessed 21 Feb 2017
5. Ordem dos Engenheiros de Portugal (2016) Avaliação de qualidade para atribuição do selo EUR-ACE. Mestrado Integrado em Engenharia Mecânica da Faculdade de Engenharia da Universidade do Porto
6. https://sigarra.up.pt/feup/pt/web_base.gera_pagina?P_pagina=257769. Accessed 21 Feb 2017
7. CDIO, Conceive Design Implement Operate. http://www.cdio.org. Accessed 21 Feb 2017
8. ENAEE, European Network for Accreditation of Engineering Education. https://www.engc. org.uk/education-skills/accreditation-of-higher-education-programmes/information-for-higher-education-providers/european-accreditation-eur-ace/applying-for-the-eur-ace-label/. Accessed 21 Feb 2017
9. Frank M, Lavy I, Elata D (2003) Implementing the project—based learning approach in an academic engineering course. Int J Techn Design Educ 13:273–288
10. Alves JL, Duarte T (2011) Research skills enhancement in future mechanical engineers. Int J Eng Pedagogy 1(1):20–26
11. Alves JL, Duarte T (2012) Short experimental ceramic projects to incentivize mechanical engineering students. Int J Eng Pedagogy 2(2):45–51
12. Lino J, Rangel B (2017) Organizational role in providing students with tools to develop professional skills (that make them employable). European WIL seminar: Work-Integrated Learning: Enable Students Pathways to Employment—Some perspectives on students, teachers and the societal role. Rectory of U. Porto, 18–19 January 2017
13. Canavarro V, Monteiro D, Rangel B, Alves JL (2017) Teaching industrial design based on real projects, a PBL experience in FEUP, Transforming Waste in industrial design products for social vulnerable groups, EDUCON 2017, Athens, Greece, April 25–28, 2017
14. Costa C, Monteiro M, Rangel B, Alves JL (2017) Industrial and natural waste transformed into raw material. Proceed Inst Mech Eng Part L: J Mats Design Appl 231(1–2):247–256
15. Aguiar C, Lino J, de Carvalho X, Marques AT (2012) Teaching industrial design at FEUP. In IDEMI 2012 Projeto Centrado no Usuário, II Conferência Internacional de Design, Engenharia e Gestão para a Inovação, Florianópolis, Santa Catarina, Brazil, October 21–23, 2012
16. Panthalookaran V, Binu R (2010) Some models and methods to nurture general management skills in engineering students living in large residential communities: ESDA 2010—ASME 2010 10th biennial conference on engineering systems design and analysis, session on Science, Engineering and Education, Istanbul, Turkey, July 12–14, 2010
17. Kostal P, Mudrikova A, Caganova D (2010) The virtual laboratory of program control. In ESDA 2010—ASME 2010 10th biennial conference on engineering systems design and analysis, session on Science, Engineering and Education, Istanbul, Turkey, July 12–14, 2010
18. Pérez CG, García PM, López JS (2011) Project-based learning experience on data structures course. In 2011 IEEE global engineering education conference (EDUCON)—Learning environments and ecosystems in engineering education. April 4–6, 2011, Amman, Jordan, 561-566
19. Zhou Z (2010) Work in progress—project-based learning in manufacturing process. In Paper presented at 4th ASEE/IEEE frontiers in education conference, session T1 J-1-2, Washington DC, October 27–30, 2010
20. Bravo E, Amante B, Simo P, Enache M, Fernandez V (2011) Video as a new teaching tool to increase student motivation. In 2011 IEEE global engineering education conference (EDUCON)—learning environments and ecosystems in engineering education, April 4–6, 2011, Amman, Jordan, pp 638–642
21. Education and Training (2017) http://ec.europa.eu/education/lifelong-learning-policy/doc48_en.htm. Accessed 21 Feb 2017
22. CEMUP (2017) http://www.cemup.up.pt/. Accessed 21 Feb 2017
23. GRANTA (2017) http://www.grantadesign.com/. Accessed 21 Feb 2017
24. Non metallic materials (2017) https://sigarra.up.pt/feup/en/UCURR_GERAL.FICHA_UC_VIEW?pv_ocorrencia_id=381005. Accessed 21 Feb 2017

25. Duarte TP, Lino J, Neves P, Araújo AC, Marques AT (2011) Materiais de construção mecânica II—uma experiência pedagógica pós-Bolonha. In CIBEM 10—X Congresso Ibero-Americano em Engenharia Mecânica, FEUP, Porto, Portugal, September 4–7, 2011
26. Metallic materials (2017) https://sigarra.up.pt/feup/en/UCURR_GERAL.FICHA_UC_VIEW?pv_ocorrencia_id=381000. Accessed 21 Feb 2017
27. Alves JL, Figueiredo MV (2016) Experimental classes of metallic materials—challenges in identifying steel components. In Paper presented at CISPEE 2016—2nd International conference of the portuguese society for engineering education, UTAD, Vila Real, Portugal, October 20–21, 2016, available through IEEE Xplore Digital Library
28. ASM, ASM Handbook (1992) Metallography and microstructures, vol 9. ASM International, February 1992
29. ASM (1999) Stainless steels. ASM Specialty Handbook, ASM International
30. Soares P (1992) Aços, características tratamentos, 5th edn. Livraria Livroluz, Porto
31. Krauss G (1980) Principles of heat treatment of steels. ASM
32. Wegst CW (2012) Stahlschussel: verlag stahlschussel wegst GMBH
33. Samuels LE (2003) Light microscopy of carbon steels. ASM International
34. IRSID (1974) Courbes de transformation des aciers. L'Institut de Recherches de la Sidérurgie Française
35. M. Atkins (1980) Atlas of continuous cooling transformation diagrams for engineering steels. ASM
36. Ramada catalogue (2017) http://www.ramada.pt/pt/. Accessed 21 Feb 2017

Jorge Lino Alves is a researcher at INEGI/LAETA and has been an associate professor at the University of Porto (DEMec/FEUP/UPorto) since 2004, in Porto, Portugal, where he teaches courses about Materials and Industrial Design. He is Adjunct Director of the Master Program in Product and Industrial Design, Director of DESIGNSTUDIO FEUP and Vice-President of the Portuguese Society of Materials (SPM). He is an integrated member of INEGI/LAETA (Institute of Science and Innovation in Mechanical and Industrial Engineering/Associated Laboratory for Energy, Transports and Aeronautics). His main research interests are additive manufacturing, industrial design, materials, new technological processes, and new methodologies in teaching engineering.

Teresa P. Duarte is a researcher at INEGI/LAETA and has been an assistant professor at the University of Porto (DEMec/FEUP/UPorto) since 1990, in Porto, Portugal, where she teaches Materials Science and Engineering and Non Metallic Materials and is responsible for dissemination and integration activities for new students at the Master Integrated Course in Mechanical Engineering. She is an integrated member of INEGI/LAETA (Institute of Science and Innovation in Mechanical and Industrial Engineering/Associated Laboratory for Energy, Transports and Aeronautics). Her main research interests are materials, new technological processes and new methodologies in teaching engineering.

António Torres Marques was born in Porto, Portugal, on September 12th 1950. He holds a second-cycle degree in Mechanical Engineering from the University of Porto (Licenciado), 1972, a MSC in Polymers, 1977, and a Ph.D. in Composite Materials from the Cranfield Institute of Technology (UK), 1981. Since 2001, holds the qualification title of 'Agregado' (habilitation) from the Faculty of Engineering, University of Porto, where he is, since February 2002, Professor at the Department of Mechanical Engineering. His areas of interest, both in research and teaching, are in Polymeric and Composite Materials, Industrial Design, Biomechanics, Health and Safety. His research activity is currently carried out within LAETA—Laboratory for Energy, Transport and Aeronautics, Research Unit of the Faculty of Engineering/INEGI. He has been responsible for several national and international projects, as well as projects with industry. Some of the projects are related to product development and include strategies for sustainable products. He has been supervisor/co-supervisor of more than 25 Ph.D. theses. He is co-author of 3 patents e published more than 320 papers in International Conferences and more than 140 in International Scientific Journals. Published, also as co-author, 12 chapters of books. He is member of Commission for Sustainability of FEUP.

"Learning by Doing" Integrated Project Design in a Master Program on Product and Industrial Design

Ângela Gomes, Bárbara Rangel, Vitor Carneiro and Jorge Lino

Abstract The Master in Product and Industrial Design (MDIP) of the University of Porto, hosted by the Faculty of Fine Arts (FBAUP) and the Faculty of Engineering (FEUP), has in its genetic code the project-based learning model. Giving the students a design studio scenario, the curriculum is developed under the integrated project design thinking, taking advantage of the knowledge provided by the two scientific areas. In a straight connection with the industry, the projects are developed in a real context, for real clients thus simulating all the tasks and stages undertaken in a design company. At the end of each exercise, the best students' concepts are developed together with the industry in response to the market need, which is a job experience opportunity in the partner company. This "formula" has been a key factor for the success of both the course and the students' career. They have the opportunity to see their project executed and implemented in the market, as well as the chance for a job opportunity in their future. In this chapter, the methodology is presented followed by the course and one example of these projects: the development of school furniture and technologies for an education company, Nautilus; this project entailed the development of a low-cost stackable and evolutionary school chair for children between 6 and 10 years old.

Keywords Integrated project design · Industrial design · Project-based learning

Â. Gomes · B. Rangel (✉) · V. Carneiro · J. Lino
Mestrado em Design Industrial e de Produto, Design Studio FEUP,
Rua Doutor Roberto Frias s/n, 4200-465 Porto, Portugal
e-mail: brangel@fe.up.pt

J. Lino
e-mail: falves@fe.up.pt

© Springer Nature Singapore Pte Ltd. 2018
M. M. Nascimento et al. (eds.), *Contributions to Higher Engineering Education*,
https://doi.org/10.1007/978-981-10-8917-6_5

1 Introduction—Integrated Project Design

The growing expertise generated by the constant advances in science and technology is, without doubt, an asset [1]. However, this has led man to perceive the world as disconnected parts [2], where knowing a lot of a small fraction is the result of a shattered intelligence [3]. The expert possesses a more in-depth knowledge about less subjects [3] whereas, in turn, the designer, such as the architect, has a wider knowledge about the problem, looking for an integrated answer [4].

In the attempt to understand design, common sense invokes the preconception that it is only restricted to aesthetics, the image [5], as if it were just a simple exercise of makeup. Design, in its essence, is interdisciplinary by nature [3, 5]. This interdisciplinary vocation is present since it works together with other disciplines/areas of knowledge during design activity, which involves, beyond doubt, very different areas of knowledge and multiple socio-technological dimensions [6–8]. Therefore, it is not surprising that a designer wanders through areas of knowledge that, at a first glimpse, would not concern him [3].

Design should not be characterized as a single discipline, but as one that can be enhanced through a variety of experiences for a wide and interdisciplinary understanding [5]. The interdisciplinary nature suggests doing something that cannot be done individually nor initiated by a single subject [9]. Unlike the multidisciplinary, where multiple disciplines are employed both in a sequential or juxtaposed mode [3, 5, 10], the interdisciplinarity aims to ensure the construction of knowledge through the transference of methods from one discipline to another [10] and, as a last resort, to break the boundaries among disciplines [11], where integration and interaction among the various areas are necessary and desirable [3, 11, 12]. This is in line with the holistic concept in which the knowledge is considered as a whole [3, 11] and where the whole is more important than the sum of its parts. Thus, the contribution of various disciplines is highly valued, indicating design as an interdisciplinary area [5] where the projects claim an interdisciplinary vision [3].

The way how products are produced has been constantly evolving, and in the recent decades, there has been a rapid growth [13], in part by consumer demand for products with better quality, lower price, better performance, and smaller delivery deadlines [14, 15]. As a result, the market requires changes in how industrial designers, engineers, and production specialists develop products [14]. Design is more and more the interface among distinct areas, attenuating the frontiers of knowledge [12], requiring a focus on research and interdisciplinary interaction through teams composed by individuals from various disciplines/areas, enhancing the lateral thinking and new methodologies more appropriate to the context in which they live. Thus, crossing information among the various areas is fundamental and inevitable for a coherent and integrated answer [1, 13]. Besides, interdisciplinarity does not mean the refusal of specialization, but rather a constant questioning to the knowledge established by it [3].

Like the response to the current requirements imposes an increasing specialization in various areas [1], new production technologies have been emerging such as additive manufacturing and new CAD software, supporting the design process and increasing the productivity, while also improving the procedures for validation and optimization of digital processes through simulation and physical models for testing and validation of concepts [16, 17]. However, they are just tools requiring a methodology to aggregate them, such as the integrated project, in the same way that in construction cement is the unifying element.

Design is not, and should not be, a simple cosmetic exercise or an individual and isolated exercise, because its interdisciplinary nature suggests doing something that cannot be done individually and is not initiated by a single subject [9]. Therefore, besides considering the user's needs and constraints of the project, it is up to the design and designer the implementation of project methodologies and knowledge from other areas, such as ergonomics, anthropology, mechanical and materials engineering, semiotics, simultaneously leading the product to a better result [3, 5, 13]. This should be always done accompanied by a team composed by the most diverse areas, promoting an integrated project. Design must always be guided by a holistic and integrative vision [1, 3, 18, 19].

To sum up, it is possible to ask what can we gain from interdisciplinarity in design?

(a) Collaboration among different areas;
(b) Acquiring new knowledge through the intersection of knowledge among areas/ disciplines;
(c) Inclusiveness, all have a vital role and something to say;
(d) Dealing with uncertainty, by avoiding take decisions based on wrong or incomplete information;
(e) Definition and framing of the problems. Most problems can only be understood when seen on common panorama among various areas [10, 18].

The goal of the integrated project is to transform a concept into a product in such way that the products design and the results of the corresponding processes evolve to minimum cost and high profitability quality product, shorter introduction of the product on the market, and lower cost of development [13, 20], responding to the new needs of consumers/users in a systematic and cohesive way, based on information from various areas. This breaking of barriers, for example, between design and production, is the fundamental key of this methodology, whose benefits include accelerating the resolution of problems during the project, where potential problems and bottlenecks are early identified and possible delays are addressed [21].

Despite the uncertainty about the impact of its applicability, many companies and sectors have successfully implemented the integrated project since its beginning in the 1990s [22], verifying, however, resistance, and reluctance to implement it [18]. At IDEO, an international company of design and innovation consulting, and a pioneer in the concurrent engineering in design [23], the myth of the solitary genius affects the company's efforts in innovation and creativity. The teams, which are at the heart of the entire process, are composed by elements of several different areas such as electrical and mechanical engineering, industrial design, ergonomics, cognitive psychology, and information technologies, which, together, are working for the same goal. At Virgin Atlantic Airways, the development phase of the project involves a series of meetings with the manufacturers to present the project and have their feedback. At Whirlpool, the innovation process and product development start on Platform Studio, where designers, experts of advanced production processes, and engineers work together to reflect on new trends and products, ending with a prototype for testing with users. At Xerox, designers, despite their experience in production, evaluate along with other specialists what is possible from an engineering and development perspective. It is still common for designers to follow engineers in customer visits to observe how these interact with the product during its use [21].

Concurrent engineering [14, 22–24], collaborative engineering [25, 26], collaborative design [17], collaborative engineering design [27], integrated design process [13, 18], integrated product development [24], and integrated project delivery [28] are some of the designations assigned to the same methodology. Despite the different names, they all have the same goal: the search for coherent solutions through interdisciplinary teams, which requires everyone to work compulsorily together from an early stage, in a constant and inclusive dialogue. It is like an orchestra where everyone is focused and linked to a shared goal [29]. No one can be excluded and everyone speaks a common language, regardless of their own language [13].

According to Dekkers et al. [22], the research on this topic—integrated project/process—highlights the importance of coordination and interdisciplinary collaboration and the advantages associated were well understood, both in academic literature and in practice, demonstrating a direct and positive effect on product innovation. The role of CAD software is extremely important and well recognized in the implementation of the integrated project [15, 22, 23, 30, 31], since without it the management capabilities of a large amount of data and information would be strangled and difficult.

Unlike traditional development processes, more time is allocated in the initial phase of the project to avoid correction of mistaken assumptions at a later stage of the process, where the opportunity to make changes decreases significantly and costs for changes increase exponentially with the advancement of the process [13, 18, 32] (Fig. 1).

Although there is no single definition for integrated project, this concept differs in intention and emphasis [18] from the conventional design process in the following aspects, which can be an asset to the project methodology in design:

Goal-driven: The goals and objectives are defined as a means to an end where those involved must demonstrate commitment instead of compliance [13, 18];
Clear Decision Making: Problem solving and decision making are based on information from different sources and areas [13, 18];
Team leader: someone responsible for the design process [13, 31];

Fig. 1 Importance of decisions in the earlier stages of product development (adapted from [32])

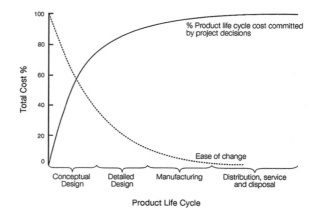

Inclusive/Collaborative: Everyone, since the client to the operator, has something important to contribute to the improvement of the function and/or performance of the product. The designer is not the "form-giver" [18], but an active participant in exploring ideas within an interdisciplinary team where everyone plays an active role since the beginning of the process and where the trust is a fundamental pillar [1, 18, 33];

Integrated: The holistic thinking is a constant where the whole is greater than the sum of its parts [1, 4, 18]. The isolated development of the components leads to worse results for the entire system, because they tend to work against each other [19];

Interactive: On the integrated project design phases are cyclical and interactive, and not linearly and sequentially as in the traditional design process (Fig. 2);

Competitive Advantage: Once it is capable of producing products with better quality and lower costs, it will be ahead of the competition.

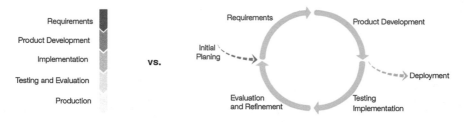

Fig. 2 Traditional sequential development method (left) versus integrated project method (right)

2   Interdisciplinarity in Design

Interdisciplinarity is defined as a methodology of knowledge integration from two or more disciplines, and depending on the capacity for dialogue and exchange between different teams, so that the work of each one is mutually enriched by the other [3, 10]. On this basis, the decision process is constantly fed by the validation of other areas, thus allowing to deepen the level of detail of the project within each area. Contrary to usual project methodologies, multidisciplinary would rule out the idea of a fragmented knowledge, scattered into diverse areas, which often prevents a link between parts and the whole [3, 33]. To achieve that complicity in different fields of study, the decisions are complemented not only by the identification of knowledge from other disciplines but also through ownership of such knowledge, i.e., through mutual combination of such knowledge thereby taking joint decisions that are built on sound technological basis [3, 33].

As Couto, cited by Fontoura [3], asserts, interdisciplinary implies "(…) a change of attitude, which allows the individual to the limits of own knowledge in order to be receptive to contributions from other disciplines. Interdisciplinarity must therefore be understood primarily as an attitude, driven by a rupture with the positivistic fragmentation-based approach, with a view to ensure a broader under-standing of reality. This approach only facilitates an effective interaction that is considered a synonym of interdisciplinarity."

Reginaldo and Baldessar [5] consider design as a field of study "without specific boundaries or defined area" and "buildup of knowledge and skills borrowed from different fields, and using a series of flexible and adjustable models that are applicable to any time and circumstances," thus in need to constantly seek innovative design methodologies. Due to the fact that it produces and applies knowledge [3], the field of design is bound to acquire knowledge specific to other areas. Engineering, ergonomics, anthropometry, and material science altogether [3, 5, 34] contribute to solving a problem, provide the answer to a specific question, or contribute to generating new ideas [10]. Nowadays, designers try to integrate the scientific methods of those areas into the design development process [35].

Globalization of markets intensifies competition and puts pressure on designers to adhere to interdisciplinarity methods as, in such a competitive context, it is no longer possible to use solely the traditional subjective and emotional methods of design [35, 36]. Currently, it is impossible to design in isolation, as no individual knows enough about the relevant disciplines that make a project a success [35, 36]. According to Bürdek [35], referring Lutz Göbel (1992) "(…) companies increasingly need neither specialists (people who know a lot about a little), nor generalists (people who know a little about a lot) but rather integralists (people who have a good overview of various disciplines with deeper knowledge in at least one area). These people must be especially capable of thinking about and acting on issues in their entirety."

According to Fontoura [3], this implies an interaction between concepts and methodologies. The interaction of concepts or the reciprocal exchange of practical and theoretical knowledge between ranges of disciplines is the basis for the area of design. The design process may be influenced by several external factors, such as the market, technologies, investment, environment, thus resulting in interventions from different fields in the development process of the project, as explained by Ashby and Johnson [37]. Ulrich (2010) referred by Bleuzé et al. [38] defines, "(…) the architecture of an artefact is more precisely as (1) the arrangement of functional elements; (2) the mapping from functional elements to components; and (3) the specification of the interfaces among interacting components."

The interaction of concepts or the reciprocal exchange of practical and theoretical knowledge between ranges of disciplines is the basis for the area of design. The design process may be influenced by several external factors, such as the market, technologies, investment, environment, etc, resulting in interventions from different fields in the development process of the project, as explained by [39]. In order to create a product, appropriate choice of materials has to be made; hence, knowledge in the area of material sciences is required [37]. The development of a product may benefit from different areas of engineering, such as chemical, electrical, production, food, or mechanical. This interaction between engineering and design will allow for better results in terms of mechanical and operational performance, costs, and durability of products [6, 34, 37].

The methodological interaction combines approaches of diverse disciplines, thus creating a common strategy to achieve more comprehensive and rigorous final results. However, to ensure the methodological or conceptual interaction, all actors in the process shall articulate their own work processes [34, 37, 40–43]. Any project development process, in any area, includes three main stages:

Definition of the problem: describes the purpose and the main objectives to be achieved [37], by creating a new product that matches the specific needs of the users and brings advantages as compared to existing competing products [42];

Definition of concepts: Ideas are proposed to meet the objectives taking into account the technical and aesthetic requirements [37];

Development: At this stage, the project is developed, from its initiation [42] and design of specifications for each component, to testing of the various components, in order to optimize the product as a whole, increasing their performance and analyzing the underlying costs [37].

Cross [34] also defines this method as a heuristic process, since designers use previous experience, general guidelines, and golden rules to define the most appropriate direction, despite not guaranteeing its success.

Currently, the process of design follows sequential tasks carried out with various tools [40, 41, 43] and allows the designer to pose questions and to seek the solutions to the problems encountered [6, 40, 44]. As it is a linear process that only allows to proceed into the next step when the previous one is concluded, is often necessary to go back to solve problems not initially detected. For this reason, some authors define the process of product development not just as a linear method but also as a straight-line method with iterative cycles [6, 40, 41]. The effectiveness of these iterative cycles is reinforced by common methodologies adopted by actors from the different disciplines. For instance, SolidWorks is used by designers but also by mechanical engineers, who use this program for numerical modeling of shapes. Also, designers use CES EduPack (Granta Design, Cambridge) to select the most appropriate materials in a sustained, technical and scientific manner. Engineering defines the concepts by analogy to previous cases, since it concentrates mainly on the functioning of products, by giving emphasis to the effectiveness of the mechanism adopted [36, 45–48].

The principle adopted is based on the use of a tool of creativity, usually designed as approach called by analogy or "design fixation," and often used intuitively by engineers and by designers [48, 49]. The analogies approach is based on the parallelism established between two products from different domains [46, 47, 50], in an explicit or implicit manner that may facilitate the adaptation or creation of new products [48, 49]. As stated by Evans [51] "design, a human activity, is discovery; it is discovery of existing but as yet undiscovered ideas."

The analogies approach allows applying the existing knowledge in a different context, thus improving the quality of the proposed solutions [46, 47, 52]. This approach is consistently based on the relational and functional similarities between the product source and the objective [40] and may take various forms. Also, analogies may be observed directly when looking into comparable situations, when the person integrates the problem while looking for the solution, when using natural elements that are similar to the problem to solve, and finally using fantasy or imagination to solve the problem as a fairy tale. Designers and engineers make use of analogies when creating new products. To invent the concept of desktop, Steve Jobs made a direct analogy with his desk where he had access to the bin, the folders, and documents. For instance, Word is similar to handwriting in a blank page where words are added in order to produce a text [50].

3 Integrated Project Design as Methodology in MDIP, a Case of Complicity Between University and Industry

Universities finally realized that it is important to train professionals instead of just researchers. When they finish their carrier, most of them enter the labor market, and only few stay in the university or in research centers. During their training, it is important to give the students this perception. Working in real context is, therefore, fundamental to train to the labor market needs [53].

Nowadays, the complex industry panorama confirms the need for interdisciplinarity, at a professional and scientific level. Efficiency of the production processes and the performance of the products are demanded in all areas of activity. Knowledge is no longer organized into different shelves, and all disciplines have to contribute to the efficiency of the product. The team is getting larger: Besides designers and engineers, end users, and industrial technicians are fundamental to optimize all the product development. The need to open the academy to the industry reality has been changing the teaching pedagogy in some courses related to engineering [54–57]. In Civil Engineering courses at the same Faculty, this experience has been led with second-year students [58]. Bringing this reality to the university has been a priority in MDIP, at FEUP, developing projects in a real context. Learning by doing is a methodology where the participants, students, teachers, and "clients" discover new paths to achieve a solution. Introducing concrete cases with real "clients" intends to establish a relation between these two worlds, allowing the student to perceive the problems in the "reality" [6]. Students are given the opportunity to communicate directly with companies and understand the production systems. Companies can develop ideas of problems, which is not possible in the daily life. This direct relation with real problems results in extra motivation and dedication to the possibility of job offers or having the projects implemented in the market.

The Master Program in Product and Industrial Design (MDIP) of University of Porto is a project-based learning (PBL) course. The training is based on unifying Courses of Project, where students, under the workshop frame (extended stay in school that can go up to 40 h/week, between tutorial hours and individual work) successively develop several products, a process in which other teachers from converging courses also participate, reflecting the aspects that relate to their subjects. The evaluation of these courses takes into account a portion resulting from the application of knowledge acquired to the projects developed [54].

Each project has the cooperation of a partner company or other external entity that launches, monitors, and validates the obtained results, without participating in the assessment, which is sole responsibility of the Faculties.

This commitment stems from the understanding that the design of new products, from capital goods to consumption ones, the so-called tradable goods, is a cooperative project-oriented business that must emerge from the intersection of three cultures: engineering, design, and management. Figure 3 shows this cooperative vision of the course [54].

Students visit companies' facilities and have contact with manufacturing processes and installed capacities. The companies present them their market and more relevant needs. During the project (held at FEUP and FBAUP), companies' technicians monitored, twice, the results under development. At the end of each semester, a public presentation of the work is made, in some cases in the partner company. If the company is interested in one or more projects, an agreement is made to the industrialization of the products.

In MDIP, this connection between the reality of the labor market and the scientific research has been the motor of its success. The practical cases are developed with the industry in a partnership association.

The students that start this course are, in its majority, designers trained in bachelor courses, and some mechanical, electrical or industrial engineers, or architects trained in integrated masters courses. In their previous courses, they were used to work with projects, but most of them without a specific client and with no possibility to be realized. Each year, in the first day meetings, new students refer

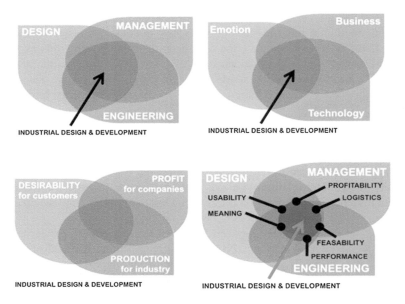

Fig. 3 Overview of cooperative industrial design and development [54]

that the choice of coming to this course is due to the possibility of having their projects realized as well as with the job opportunity that this methodology offers. Since the beginning, students know that their effort and their professional capacity will be fundamental to achieve their expectations.

The scenario that is presented to the students is equal the one that they would have in a design studio, or an industry. They are organized into teams to develop different solutions for the same program that will compete in the final presentation for the client.

The fact this course is a partnership of both the Faculty of Engineering and the Faculty of Fine Arts provides the students with different approaches and makes them understand the meaning of integrated project design (IPD). The different disciplines feed the project that is developed in the design studio disciplines. Once again, the real scenario is presented; they search for the technological answers in the related discipline with the specialized teacher.

After the first year, students can develop the selected projects by the client and the teachers, as a research problem for their one-year master thesis. They are asked to improve their projects making a scientific research, increasing the state of art related to the technical and scientific issues and performing a set of experiments to find the solution to industrialize the concept they have created in the first year. In this second year, the scenario is close to that of an industry research center in the University workshops and laboratories. Like the professional practice, they prepare their projects in an integrated project design (IPD) thinking, developed with teachers with different backgrounds, designers, architects and mechanical engineers. After this phase of research, students go to the companies or industries to improve the project in a professional stage program, until its industrialization.

Some of the projects developed in the past years are now in the industries and in the commercial market. A glove for motorcyclists is now in the final phase of the industrial process. An evolutionary chair for children between 6 and 10 years old is now in the market. The primary schools from the municipality of Penafiel are today using the Dual-Step chair. The research made, searching for a universal chair, was recently published in Applied Ergonomics [59], which demonstrates the quality of the work that has been done.

After these two years, students realize the different tasks involved in the career that they chose, in both the professional or scientific paths. They learn, by doing, that all the disciplines involved in the production of an item, have to work together to achieve an integrated solution that responds to nowadays strict requirements.

4 Case Study—Developing an Evolutionary Chair for Children Between 6 and 10 Years Old

In this chapter, an example of these exercises, developed in the two years of the course with Nautilus Company, is presented; the development of an evolutionary chair for children between 6 and 10 years old. The concept was created during the first year in a teamwork and in straight collaboration with the company. During the second year, after a deep scientific research on the ergonomics issues involved, two of the students had a professional experience in the company, developing the projects in industrial context. One of those chairs is now in the company catalogue and already in use in some primary schools. At this moment, one of the students is already part of the company staff.

Like other companies that collaborate with MDIP, Nautilus, a Portuguese company specialized in school furniture and education technology, asked in the first meeting with the teachers to solve a problem that they had for a long time. The proposal was the development of a school chair for primary education, particularly for children from 6 to 10 years old, which should be adjustable in height, stackable and with a low production cost. It was a problem that they wanted to solve, but in the day-by-day life there was never time to think about it. Finding a solution for a universal chair would deeply optimize all the company chairs manufacturing process. There would be just one size of chair, one cast in the production, one type of stock, one type of order.

In the first year, the concept was defined in teams with distinct backgrounds, designers, and mechanical engineers. During the first week, the students visited the different factories of the company and perceived the available technologies. Similar to a design studio context, the teams proposed different ideas to the same problem. After some weeks of work in the Design Studio of FEUP, the first ideas were presented to the company to validate the concept proposed. In the second part of the exercise, each student should develop the concept in detail until the execution project scale. At this phase, being in the Faculty of Engineering was crucial. The students easily found the information on the specifics issues that were arising with the teachers of different areas, design, ergonomics, mechanical engineering, etc. Although they were developing individual proposals, they worked as a team, helping each other. The mechanical engineer students helped the designers to study the mechanical behavior of the solutions in SolidWorks. The designers helped the mechanical engineers to prepare their presentations with InDesign or Photoshop. The students with an Engineering background helped in the definition of the mechanical details. The students with a design background helped in the aesthetic of the product. It was a real collaborative work, like in the integrated project teams. Meanwhile, when they had specific problems related to the manu-facturing process of their solutions, the company was always available to help.

At the end of this first year, the final presentation was made for the company. Students prepared their individual projects as if they were in a competition. The company would choose one or two ideas to be industrially developed in the fol-lowing year. The presentations were very professional, each one made a short movie of the proposal, explaining the production system and the use of the product, a poster communicating the idea, a marketing flyer, a folder with the drawings details and specific information, besides a small-scale 3D-printed chair prototype. The CEO of the company was surprised with the quality of proposals and instead of choosing one he selected two proposals.

In the second year, two students developed the selected projects. In a first phase in the Faculty, exploring the ergonomic issues concerned with the universal chair, and in a second phase redesigning their projects to make them industrialized. A detailed research on the design of the universal chair was conducted with the teachers [59].

After two months, the students went four days a week to work in the company, and once a week they met with the teachers to report the development of the project. They went through the various phases of production; the redesign, the preparation of production, the production set, the validation, and the implementation. Now, they can see one of the products being used by primary school students in a small city close to Porto and the product already in the catalogue of the company.

In this chapter the result of three of these projects is presented, two of them developed in the second phase of the exercise, with the company in an industrial context.

4.1   Three Concepts, Three Solutions, Two Research Problems

Of all the projects developed, three adopted the integrated project as the working methodology. Unlike a current design process, the starting point for the solution to be developed was defined by the mechanical behavior of the chair and regulation system. The projects were of high technical and scientific consistency and as a result were selected by the company to be developed in a business environment with the purpose of its production and commercialization.

In project A, from student Vitor Carneiro, the starting point was defined by the mechanical behavior of the chair, determined in part by the concept adopted, in order to develop the final solution. As such, a convergence among form, regulatory mechanism, and material was paramount and the focal point in the development of the solution, where the need to use engineering tools, such as CAD software and material selection methodologies, would allow a support and a more sustained justification of the solution adopted than only based on aesthetic and formal criteria. Throughout the entire development process, everything has been accomplished, designed, and conceived in an integrated environment where partial and complete simulations have been done continuously in order to achieve a valid and justified solution in several parameters.

Project B, developed by the student Ângela Gomes, began by exploring the mechanical response to the problem under study. Finding the concept of interdisciplinarity based on two essential premises, interaction between concepts and interaction between methodologies, engineering concepts were used, more specifically those related to the mechanisms, in order to solve the height adjustment problem of the chair, while also ensuring that it was stackable. She applied a methodology of product development learned in one of the curricular units (CU) of the course. This method consisted in the search for mechanisms of regulation, even if they were not used with the purpose of regulating height. After researching the existing mechanisms, she tried to perceive the functioning of each one, disassembling them and analyzing how the various components were related.

The methodology used in the project is based on the method of product creation adopted by engineering. This method is based on the use of a creativity tool called by analogy, usually used in an intuitive way, both by engineers and by designers [48, 49]. Analogies are based on the parallelism created between two products with different domains [46, 47, 50], allowing to apply existing knowledge to another context, thus improving the quality of solutions [46, 47, 52]. By always ensuring the principle of the relational and functional similarity between the source product and the objective [40], analogies can take several forms. After the chair was developed, a direct analogy was used, since for the generation of the mechanism a search of several existing mechanisms was made, selecting next those that would be the most suitable for the type of regulation that was intended, making adaptations of the same to the context in which they would be applied.

In project C, initially developed by student Maria João Pato and later by Vitor Carneiro, introduces a radical innovation in the market of adjustable school chairs by exploring an existing lack, but quite evidenced by the literature: the lack of regulation of the depth of the chair. As vital as the height of the chair of the chair is the depth of it, so the starting point in this project was defined by the development of a system capable of simultaneously adjusting the height and depth of the chair, without thereby compromising the stacking of the chair, and its low production cost.

In all the three projects, the starting point focused on research of the state of the art, legislation, standards, materials, adjustable mechanisms, and user needs. It was observed that:

(a) From the existing school chairs, only a tiny number conciliated the requirements of stacking and regulation of chair height;
(b) From the adjustable chairs in the market, the chair adjustment was always made step by step and mostly using the pin mechanism (Fig. 4);
(c) Stacking the chair does not have the same meaning for the companies as for the school. From the companies' point of view, stacking is about promoting less storage space and a larger number of chairs per load in transport—vertical stacking. From school point of view, vertical stacking is not a daily necessity as opposed to lateral stacking. For daily cleaning of the classrooms what matters to schools is that the chairs can be placed lying on the table and stacked laterally promoting efficiency and speed;
(d) It is very common that in the same classroom the chairs are the same size mark, which is not the best option. It was found that within the same class there can be differences of 200 mm between the lowest and the highest students;
(e) The use of school furniture not suitable for its users is enhanced for bad postures, which have a direct influence not only in its correct and normal growth, but also in their cognitive development [60–65].

Given the aforementioned observations, it was defined, as design principles, a continuous adjustment of the chair rather than step by step; the student should be able to regulate the chair itself; the chair should be stackable; and physical lightness and aesthetics should be present. Given these facts, it is more than evident the need for an integrated project where three main disciplines prevail; design, ergonomics and engineering, especially when the need to combine stacking and continuous regulation of the chair seat have made focus the development of the concept in a strategic alliance, almost like a game of balance between form and mechanism, design and engineering, finding a viable and valid solution (Fig. 5).

A continuous and simultaneous focus on the mechanism, in the shape and taking into account users' needs and requirements were essential to the project. This interaction between design and engineering, working together to achieve the

Fig. 4 Pin mechanism

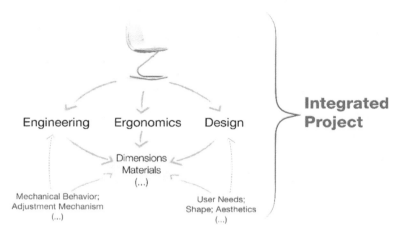

Fig. 5 Contribution of different areas for the school chair development under the integrated project methodology

optimal parameters in the definition of the product, can start from a simple communication of ideas until the interaction of concepts.

Once the project requirements were outlined, the students' proposals followed different paths, always accompanied and validated by teachers with different skills and company technicians.

4.1.1 Project A: Clamp Chair

In the pursuit for a chair that would allow a closer dimensional of the anthropometric user needs, the definition of design concept emerged from the principle of a continuous regulation and, because of usability issues, in one place which presupposed the convergence of the various support points of the chair to only one. The chair development assumed a convergence between form and mechanism. To achieve it, it was essential to understand the mechanical behavior of the chair and to focus in the solution development, where the need for engineering tools, such as CAD software, would allow a stronger justification of the adopted solution than just based on design criteria.

The most promising concept emerged from a strong mechanical component combined with mechanism's simplicity in which the materials would have a key role to play. With a Z shape and composed by two distinct parts, the structure and the shell, the adjustment is accomplished by telescopic tubes, which make up the structure, and its position lock/unlock is performed through a single system, the same used in bicycles seat, which allows a comfortable and easy handling and also a continuously adjust. Using a simple guide system it was ensured that there is no misalignment between the shell and the structure, thereby guaranteeing the permanent balance of the chair (Fig. 6).

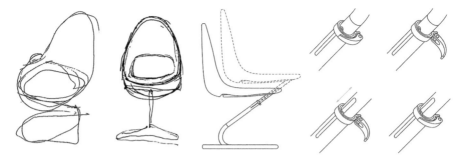

Fig. 6 Sketches of the relation between form and mechanism in the school chair design

The active participation of designers and mechanical and materials engineers (teachers, students, and company technicians) was vital to get a feedback of what be possible to achieve in terms of solutions. The need for its detail and specification, such as dimensions, materials, mechanical behavior, were evident and essential. The use of CAD software, 3D modeling, and mechanical behavior simulation were needed for a product detail.

The Right Material in the Right Place!

Nautilus uses mainly two types of materials for the development of school furniture: metal, mostly low carbon steel, and plywood. However, the thermoplastic polymers were recently included as election material, investing on the acquisition of injection molding equipment for production. Thus, and given the primary form of the concept, which divides the chair into two components, the structure and the shell, the metal and polymer were the materials family chosen for the concept development. The need to refine the choice of materials was a vehement necessity, particularly the mechanical behavior of the structure whose shape needed to be refined through simulation tests. "The right material in the right place" [24].

Screening and selection are two key steps in material's selection process [66]. The first allows a reduction to a manageable number of materials, while the second one allows obtaining a ranking of potential materials in agreement with the established requirements. Based on a systematic methodology of selection materials through a TOPSIS analysis, considered the best method of multi-attribute decision making selection [66], a ranking of potential materials for the structure and the shell was obtained taking into account the requirements and objectives defined (Fig. 7).

Delineated the materials, the analysis of the mechanical behavior of the chair when subjected to stress was performed in order to achieve a viable solution and establish dimensions such as tube thickness. Through SolidWorks software, the 3D modeling of the chair was carried out and, subsequently, the static simulation tests to the structure and the shell to make any necessary corrections and improvements. A behavioral analysis of the structure and the shell assembled was performed

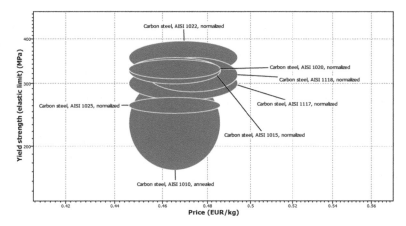

Fig. 7 Yield strength versus price for different types for construction steels (CES Edupack software)

afterward, since the behavior of the set would effectively be slightly different from the individualized behavior. The tests were performed in SolidWorks, through the simulation application (Fig. 8).

Through the improvement of the form and the constructive solutions, it was possible to decrease the nodal stress (von Mises), and resulting critical strain, in the various static simulation tests conducted to the structure. This is a cyclic cooperation between design and engineering, or "functional design and industrial design" [26], in a single design process where both evolve in a way that they are influenced by each other and in which the sketch is a constant tool for refining ideas and facilitating problem solving [6] (Fig. 9).

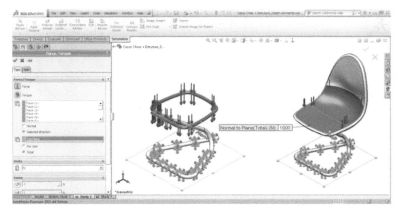

Fig. 8 Force (N) applied in the static tests

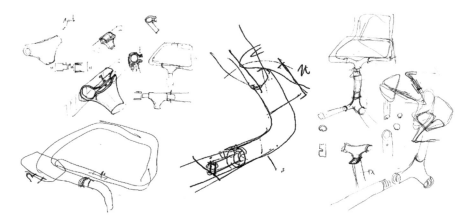

Fig. 9 Sketches for search constructive solutions and improvement

This interface between design and engineering allowed that the nodal stress in the order of 930 MPa, well above of the material yield strength, was successively decreasing until obtaining an acceptable 298 MPa. It was important to obtain the mechanical behavior of the assembled structure and shell. Everything was performed, designed and conceived in an integrated environment where partial and complete simulations were continuously made, in order to achieve a valid solution and justified on the various parameters defined during the product development phase. The result proved to be a project of high technical and scientific consistency. However, and despite the interest expressed by the company, the project turned out not to proceed to production (Fig. 10).

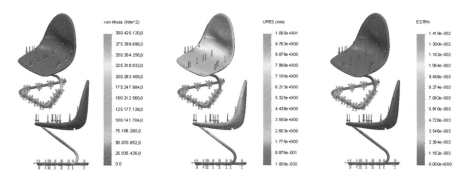

Fig. 10 Nodal stress (von Mises), displacements (URES), and equivalent strain (ESTRN) of the final structure and shell assembled (deformation scale = 1)

4.1.2 Project B: Seat and Growth Chair

One of the projects selected by the company was Seat and Growth Chair, developed by Ângela Gomes. After a deep research on the ergonomic issues concerned with the height adjustment of the seat, the student developed the prototype in the company, taking part in all the phases of the industrial process. Starting from the deep study of the mechanism, the design of the chair was developed based on solid and technologically sustainable decisions. Using a shared relationship promoted by the integrated project, it was possible to improve the product's functionalities, ascertain the quality of production, and thus enhance its creativity. Having a greater number of information about the project, the designer can solve problems in an innovative way.

The project B had as a starting point the operation of the mechanical response to the problem in study. After the research made to support the design process, the student established that to achieve the company objectives, two questions prevail:

- Is it possible to improve the usability of the adjustment mechanism without compromising its effectiveness?
- Is it possible to adapt this new mechanism to a chair developed according to the existing production processes in the company?

To answer the first question, a deep research on mechanical system was led by the student. In order to tackle the requirement that the chair would be simultaneously height-adjustable and stackable, data was gathered related to the existing mechanisms of height adjustment and the diverse stacking systems. An analysis of products for a similar purpose or products with a different purpose but similar mechanisms, offered by the main competitors in the market, was made. As the product targets the specific market niche of schools, the solution to propose should be produced and offered in the market at low costs and prices. It was therefore decided to analyze manual mechanisms of simple operation, as the use of more sophisticated technologies would increase the production costs.

Once the first drawings were finalized (Fig. 11), a few modeling tests were applied in order to check the matching of main selected concepts with the required functionalities. To validate this operation and better understand the stress resistance of the chair, the students consulted their mechanical engineering experts.

When the mechanism was selected, it was concluded that, in order to satisfy the technical requirements of the product, the concept should be submitted to additional improvements, through a systematic trial and error approach of the solutions to the problems encountered. On this basis, the iterative process resulted in the final concept design of the chair (Fig. 12).

Designed for the child to be able to adjust easily the chair, the height adjustment mechanism was inspired on the trestle system. Stacking is possible from above and up to five chairs. The seat is manufactured from wood and different colors are possible, as shown in Fig. 13. It is also suggested to apply a different color to the adjustment mechanism. The frame is manufactured from steel, selected considering its high resistance.

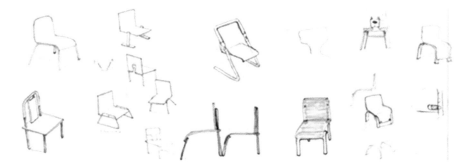

Fig. 11 First drawings

Phase 1

Modifications
— Mechanism inserted within the frame

Shortcomings
— production of the frame with inserted mechanism very expensive;
— rear of hull would not be supported by the frame

Phase 2

Modifications
— Easel mechanism inserted in the front of chair;

Shortcomings
— rear of hull would not be supported by the frame

Phase 3

Modifications
— Mechanism tested in phase 2 but support added at the rear;

Shortcomings
All tested mechanisms did not match the requirements

Phase 4

Modifications
— Trestle-type mechanism of supports, inserted in the rear and within lower frame (legs), of wood

Shortcomings
— wood is very expensive; option: use another material

Fig. 12 Project evolution (from left to right)

Fig. 13 **a** Adjustment system (inspiration); **b** stacking

(a)   **(b)**

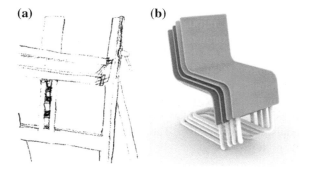

After having the concept design of the chair, the next step was to adequate this concept to the production constraints of the company. The main corrections pointed out by the company were the method of fixing the adjustment system, the dimensions of the tube utilized and the need to consider the Standard in use to scale out the chair. In the company, the design and technical team analyzed the project and suggested a redesign of the product, to be adapted to the technologies available in their factories and with regulations Standards that they work with [67].

The next step was to adequate this concept to the production constraints of the company. Once selected the material of the different components, the definition of the production processes was made. Having the processes of the company as a starting point, the structure was produced in three steps: cut off the tubes, curving the tubes using the CNC process, and welding the different components. The seat was produced by making the plywood laminate and conforming the shape in a vacuum press available on the company (Fig. 14). The last component was the regulation system for this and having in account the material, it was decided that the best process was the milling cutter.

At the end of the production process, some prototypes were made resorting in an iterative process of testing and concept improving. The validation of the proposal was incomplete due to the fact that the chair was not completely finished, requiring some changes and testing by children and in a school environment. Besides the proposed changes, it was still necessary to reflect on the manufacturing processes of the different components of the chair, considering the costs and production times, to optimize its production (Fig. 15).

The cooperation with Nautilus was fundamental so that the concept of Seat and Grow, developed in the curricular unit of Industrial Design Project, was transferred from theory to practice. The access to the materials, technologies and manufacturing processes of the company, allowed knowing the industrial universe, helping in the decision making during the improvement of the concept. In addition to the industrial constraints, it was also possible to realize that production times influence the evolution of the development of new products.

Seat

Fig. 14 Production of the seat

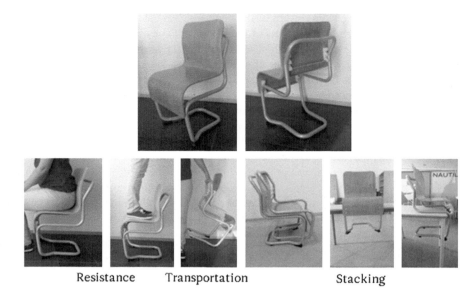

Resistance   Transportation   Stacking

Fig. 15 Last prototype and some tests that were made

4.1.3 Project C: Dual_Step Chair

The second project selected by Nautilus was the Dual_Step Chair, developed by student Maria João Pato in the first phase and Vitor Carneiro in the second one. Instead of starting from a new design for the chair, Maria João Pato chose to redesign one of the company's chairs already in catalogue, the Uni_Step chair.

The objective was to find a system that could make a three-dimensional adjustment with a simple mechanical system. The intervention of teachers of different areas was fundamental to find the answer, from architects to mechanical engineers. In the classroom, the help from the other students with different background was also crucial to find the concept of this solution. After the first year, student Vitor Carneiro developed this project in his research thesis making a study on a universal chair design system and following the industrialization of this concept.

The concept of this project appears with the analysis of the European standard for the design of school furniture, EN 1729-1: 2006 [67], and the perception of dimensions variations in sizes needed to accommodate primary school children. Between size 2 and size 4, proposed by the European standard, seat height and depth may vary by 80 mm. From these requirements, it was assumed that the height of the seat had a great impact in ergonomic terms, which would explain why the height-adjustable school seats have a fixed seat depth, as was verified in the market research carried out. However, in a height-adjustable chair with a single depth, two situations may occur. If a greater depth is used, children with a smaller

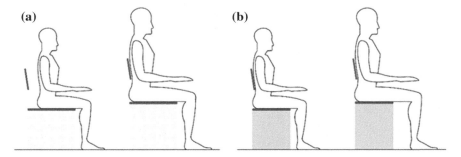

Fig. 16 Consequences of a single depth on adjustable seats: **a** adjustable seats that use a seat with greater depth means those smaller children cannot use the backrest; **b** adjustable chairs that use a seat with a lower depth leads to lack of support in the thighs in older children

gluteal-popliteal length cannot properly use the backrest (Fig. 16a). If a lower depth is used, it may lead to a lack of adequate support of the thighs in children with a greater gluteal-popliteal length (Fig. 16b).

For all this, it was decided to include another requirement for the project in order to improve the ergonomic aspects of a school seat: The height and depth of the seat should be adjustable, preferably through a single movement. The ability to change both dimensions simultaneously will allow children to learn early on what should be the correct posture when seated.

As a result of this analysis, it was necessary to move two axes simultaneously, so that small children would have the height and depth of the seat in the lowest size, and as they grow, the seat could be simultaneously adjusted to a greater height and depth (Fig. 17).

According to Zimmerman [18], in some cases the best response may not be the result of a new product, but rather add value to an existing product. In this way, and taking into account the fact that the cost of the product varies according to the materials and manufacturing processes used, it was decided to develop this chair based on the design of the Nautilus Uni_Step chair. The procedure was thus the transfer of knowledge, components, and materials (Fig. 18). With this approach,

Fig. 17 Operating diagram of the two dimensions: seat height and depth

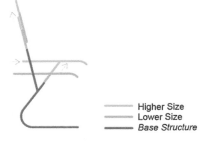

—— Higher Size
—— Lower Size
—— *Base Structure*

Fig. 18 Nautilus Uni_Step Chair (left) versus Dual_Step Chair concept (right)

one could combine the best of both worlds. The first two-dimensional adjustable school chair that allows achieving proper regulation with just one movement, and a quality product with low cost of production, therefore, accessible to schools.

Recognizing the innovative nature, clear benefits, and economic viability of the Dual_Step Chair, Nautilus has shown an interest in advancing with the project in a business context, integrating the student into its multidisciplinary development team. In the process of detailed design of the Dual_Step Chair, and in order to achieve an economically viable product that can be produced in the company, we worked on three fronts to achieve a technically and formally capable product. Technical solutions were developed for the mechanism of regulation and connection between the base structure, the seat structure, and the backrest structure. The improvement, testing, and validation of the locking handle of the chair position were sought. The redesign of the chair was carried out, according to the completely productive process of the company. Through sketching, 3D modeling software SolidWorks, and chair prototypes, various solutions have been developed, matured, and tested. Several chair prototypes were produced and subjected to several rigorous tests in order to guarantee the efficiency, effectiveness, safety, and quality of the chair until the final solution was reached (Fig. 19).

With the end of this research and development process, the Dual_Step Chair, the first school chair with adjustable seat height and depth, was ready to equip the first schools, making them an asset in promoting good posture practices at school (Fig. 20).

This contribution was evidenced by the Institute of Industrial Engineers (IIE) and GOErgo, which awarded the Dual_Step Chair with the Creativeness in Ergonomics Student of the Year Award, at the 18th Annual Applied Ergonomics Conference 2015 in Nashville (Fig. 21). This award recognizes achievements in the research and application of ergonomics, including process improvement, applied instrumentation, and product development.

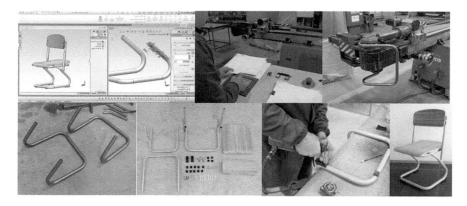

Fig. 19 Example of the production process of one of the prototypes of the Dual_Step Chair

Fig. 20 First Dual_Step
Chairs equipping a classroom
of the Irivo School Center

Fig. 21 Ceremony of
delivery of creativeness in
ergonomics student of the
year award to Maria João Pato
and Vitor Carneiro

5 Conclusions

The integrated project is a methodological approach with considerable advantages in product development, favoring an integrated environment in which individuals from different areas of knowledge contribute together to a valid product solution, justified by several criteria transversal to various areas. In product development it is unthinkable not to associate design, engineering and ergonomics. In fact, the market demands it and companies have to adapt themselves to these new demands: better products, better quality, better performance, lowest prices, and smaller delivery deadlines.

The application of this methodology in academic environment, particularly in product and industrial design courses, is a natural evolution that allows achieving a greater depth in detail, a more complete response in product development and, consequently, a greater preparation of their students to the new consumer requirements and business needs. As demonstrated by this case study, during the whole development process of the height-adjustable school chair, the cooperation between design and engineering was fundamental to the transformation of an idea into a commercial product. Still in the early stages, the relationship mechanism/form, engineering/design was evident in the evolution of the concept. It was even deeper and intrinsic in the dimensions definition, constructive solutions, and materials, where CAD software was an asset, and the exchange of concepts/knowledge among areas played a crucial role to achieve the final solution.

Based on the results of these three projects, it is possible to conclude that the integration of an interdisciplinary design methodology in the development of new products allows us to reach innovative and functional solutions, once creating a relation of sharing, both of knowledge and of methodologies, between design and engineering, there is a greater perception of the projects in question. By providing the designer with a greater number of information, his creativity is enhanced, thus solving the problems in an innovative and complete way. Combining this information with the use of creativity tools, in particular with the method of analogies, solutions to particular problems can be found in other ways.

Acknowledgements The authors gratefully acknowledge all the MDIP students that participate in this project, the funding of Project NORTE-01-0145-FEDER-000022—SciTech—Science and Technology for Competitive and Sustainable Industries, co-financed by Programa Operacional Regional do Norte (NORTE2020), through Fundo Europeu de Desenvolvimento Regional (FEDER) and Community Service Engineering European Project (LLP-539642-Community Service Engineering).

References

1. Rangel B, Alves FB (2013) Engineering as a lesson in architecture. Cadernos d'Obra 92–94
2. Behrens MA (1999) A prática pedagógica e o desafio do paradigma emergente. Revista Brasileira de Estudos Pedagógicos 80:383–403

3. Fontoura AM (2011) A interdisciplinaridade e o ensino do design. Projética Revista Científica de Design 2:86–95
4. Souto Moura E (2009) Edifício Burgo: design, construction, technologies. In: d'Obra C, Rangel B, Faria JA, Martins JPP (eds). Gequaltec, ed. Porto
5. Reginaldo T, Baldessar MJ (2013) O conhecimento disciplinar do Design e suas contribuições para a teoria interdisciplinar in Simpósio Internacional sobre Interdisciplinaridade no Ensino, na Pesquisa e na Extensão – Região Sul. Florianópolis
6. Dym CL, Agogino AM, Eris O, Frey DD, Leifer LJ (2005) Engineering design thinking, teaching, and learning. J Eng Educ 94:103–120
7. Daly SR, Adams RS, Bodner GM (2012) What does it mean to design? A qualitative investigation of design professionals experiences. J Eng Educ 101:187–219
8. Détienne F, Martin G, Lavigne E (2005) Viewpoints in co-design: a field study in concurrent engineering. Des Stud 26:215–241
9. Little A, Hoel A (2011) Interdisciplinary team teaching: an effective method to transform student attitudes. J Eff Teach 11:36–44
10. Domik G (2008) Teaching visualization in multidisciplinary, interdisciplinary or transdisciplinary mode. http://pdf.aminer.org/000/591/607/a_multi_disciplinary_look_at_the_computing_disciplines.pdf. Accessed 30 Mar 2017
11. Gadotti M (2006) Interdisciplinaridade: Atitude e Método. Instituto Paulo Freire, Ed. São Paulo, p 7
12. Chew E (2006) Imparting knowledge and skills at the forefront of interdisciplinary research—a case study on course design at the intersection of music and engineering. In: 36th ASEE/IEEE frontiers in education conference, San Diego, CA
13. Magrab EB, Gupta SK, McCluskey FP, Sandborn PA (2009) Integrated product and process design and development: the product realization process, 2nd edn. Taylor & Francis, Boca Raton
14. Xu L, Li Z, Li S, Tang F (2007) A decision support system for product design in concurrent engineering. Decis Support Syst 42:2029–2042
15. Goldin DS (1999) Tools of the future. J Eng Educ 88:31–35
16. Diefes-Dux HA, Samant C, Johnson TE, O'Connor D (2004) Kirkpatrick's level 1 evaluation of the implementation of a computer-aided process design tool in a senior-level engineering course. J Eng Educ 93:321–331
17. Shen Y, Ong SK, Nee AYC (2010) Augmented reality for collaborative product design and development. Des Stud 31:118–145
18. Zimmerman A (2006) Integrated design process guide. Ed. Ottawa: Canada Housing and Mortage Corporation, p 18
19. Hawken P, Lovins AB, Lovins LH (2000) Natural capitalism: creating the next industrial revolution. Little, Brown & Company, Boston
20. Friel T (2000) A dramatic method to demonstrate concurrent engineering in the classroom. J Eng Educ 89:265–267
21. Design Council (2007) Eleven lessons: managing design in eleven global brands. Ed. London, p 144
22. Dekkers R, Chang CM, Kreutzfeldt J (2013) The interface between "product design and engineering" and manufacturing: a review of the literature and empirical evidence. Int J Prod Econ 144:316–333
23. Thomke S, Nimgade A (2007) IDEO product development. Harvard Business School, vol. Case: 600-143
24. Ljungberg LY (2003) Materials selection and design for structural polymers. Mater Des 24:383–390
25. Willaert SSA, de Graaf R, Minderhoud S (1998) Collaborative engineering: a case study of concurrent engineering in a wider context. J Eng Technol Manag 15(1):87–109
26. Mas F, Menéndez JL, Oliva M, Ríos J (2013) Collaborative engineering: an airbus case study. Procedia Eng 63:336–345

27. Juhl J, Lindegaard H (2013) Representations and visual synthesis in engineering design. J Eng Educ 102:20–50
28. American Institute of Architects (2007) Integrated project delivery: a guide. Ed. California
29. Rocha PMD, Furtado R (2013) Museu Nacional dos Coches: design, construction, technologies. In: d'Obra C, B. Rangel, Faria JA, Martins JPP (eds) Gequaltec, ed. Porto
30. Smailagic A, Siewiorek DR, Anderson D, Kasabach C, Martin T, Stivoric J (1995) Benchmarking an interdisciplinary concurrent design methodology for electronic mechanical systems. In: 32nd ACM/IEEE design automation conference, San Francisco, 514–519
31. Iansiti M, MacCormack AD (1997) Team New Zealand (A). Harvard Business School, vol. Case: 697-040
32. Norman G (1990) Life cycle costing. Prop Manag 8:344–356
33. Borrego M, Newswander LK (2008) Characteristics of successful cross-disciplinary engineering education collaborations. J Eng Educ 97:123–134
34. Cross N (2008) Engineering design methods: strategies for product design, 4th edn. Wiley, New York
35. Bürdek BE (2005) Design history, theory and practice of product design. Birkhäuser, Basel
36. Charyton C, Jagacinski RJ, Merrill JA, Clifton W, DeDios S (2011) Assessing creativity specific to engineering with the revised creative engineering design assessment. J Eng Educ 100:778–799
37. Ashby M, Johnson K (2010) Materials and design: the art and science of material selection in product design, 2nd edn. Butterworth-Heinemann, Oxford
38. Bleuzé T, Ciocci M-C, Detand J, De Baets P (2014) Engineering meets creativity: a study on a creative tool to design new connections. Int J Des Creat Innov 2:203–223
39. Treball E, García JF, García VL, Viñes JV (2009) Ergonomía: diseño centrado en el usuario: Fundación Prodintec
40. Crul MRM, Diehl JC. Netherlands. Delft University of Technology. Faculty of Industrial Design Engineering, Design for sustainability (D4S): a step-by-step approach. United Nations Environment Program, Paris
41. Parsons T (2009) Thinking: objects—contemporary approaches to product design. AVA Publishing SA, Suiça
42. García JF, García VL, Santacoloma S (2006) Diseño industrial: guía metodológica: Fundación Prodintec
43. Ullman DG (2010) The mechanical design process, 4 ed. McGraw-Hill Higher Education, Boston
44. Curry T (2014) A theoretical basis for recommending the use of design methodologies as teaching strategies in the design studio. Des Stud 35:632–646
45. Buckland D (2012). How physicians, engineers, and scientists approach problems differently. http://www.medgadget.com/2012/08/how-physicians-engineers-and-scientists-approach-problems-differently.html. Accessed 30 Mar 2017
46. Casakin H, Goldschmidt G (1999) Expertise and the use of visual analogy: implications for design education. Des Stud 20:153–175
47. Moreno DP, Hernández AA, Yang MC, Otto KN, Hölttä-Otto K, Linsey JS et al (2014) Fundamental studies in design-by-analogy: a focus on domain-knowledge experts and applications to transactional design problems. Des Stud 35:232–272
48. Robinson JA (1998) Engineering thinking and rhetoric. J Eng Educ 87:227–229
49. Crilly N (2015) Fixation and creativity in concept development: the attitudes and practices of expert designers. Des Stud 38:54–91
50. Zax D (2014) How Steve Jobs' mastery of analogies sent Apple skyrocketing. http://www.fastcompany.com/3037014/my-creative-life/how-steve-jobss-mastery-of-analogies-sent-apple-sky-rocketing. Accessed 30 Mar 2017
51. Evans R (2003) Design design: a theory of design. Int J Eng Educ 19:81–93
52. Baldaia J (2010) Emoções, analogias e Criatividade, INTUINOVARE vol 2014, J. Baldaia, Ed.

53. Reigeluth CM (2009) Instructional-design theories and models: a new paradigm of instructional theory, vol 2, 1st edn. Lawrence Erlbaum Associates, New York
54. Aguiar C, Lino J, Carvalho X, Marques AT (2012) Teaching industrial design at FEUP. Presented at the IDEMI 2012—II Conferência Internacional de Design, Engenharia e Gestão para a inovação, Florianópolis, SC, Brasil
55. Frank M, Lavy I, Elata D (2003) Implementing the project-based learning approach in an academic engineering course. Int J Technol Des Educ 13:273–288
56. Martinez L, Romero G, Marquez JJ, Perez JM (2010) Integrating teams in multidisciplinary project based learning in mechanical engineering. In: IEEE EDUCON 2010 conference, pp 709–715
57. Zhou Z, Donaldson A (2010) Work in progress—project-based learning in manufacturing process. In: 40th IEEE frontiers in education conference (FIE)
58. Rangel B, Guimarães AS, Sá AV, Alves FB (2016) Integrated design concept in civil engineering education. Int J Technol Des Educ 32:1279–1288
59. Carneiro V, Gomes Â, Rangel B (2017) Proposal for a universal measurement system for school chairs and desks for children from 6 to 10 years old. Appl Ergon 58:372–385
60. Gouvali MK, Boudolos K (2006) Match between school furniture dimensions and children's anthropometry. Appl Ergon 37:765–773
61. Panagiotopoulou G, Christoulas K, Papanckolaou A, Mandroukas K (2004) Classroom furniture dimensions and anthropometric measures in primary school. Appl Ergon 35:121–128
62. Parcells C, Stommel M, Hubbard RP (1999) Mismatch of classroom furniture and student body dimensions: empirical findings and health implications. J Adolesc Health 24:265–273
63. Reis PF, Reis DCd, Moro ARP (2005) Mobiliário Escolar: Antropometria e Ergonomia da Postura Sentada. Presented at the XI Congresso Brasileiro de Biomecânica, S. João, PB
64. Moro ARP (2005) Ergonomia da sala de aula: constrangimentos posturais impostos pelo mobiliário escolar. Rev Digit 10:1–6
65. Castellucci HI, Arezes PM, Molenbroek JFM (2014) Applying different equations to evaluate the level of mismatch between students and school furniture. Appl Ergon 45:1123–1132
66. Jahan A, Ismail MY, Sapuan SM, Mustapha F (2010) Material screening and choosing methods—a review. Mater Des 31:696–705
67. Rangel B, Alves FB (2013) Engineering as a lesson in architecture. Cadernos d'Obra, pp 92–94

Ângela Gomes is graduated in Technology and Product Design (2010–2013), at the Aveiro Norte University School, Aveiro University, and her first professional experience was at the end of the degree course a three-month curricular internship in the machine production, laser, cutting and milling cutters industry at Optima, Lda. belonging to the Tecmacal Group, located in São João da Madeira. At the end of her degree, she studied in the master of Industrial and Product Design (2013–2015) from the Faculty of Fine Arts and the Faculty of Engineering of the University of Porto. Her thesis was developed in a business environment, using a professional internship in the industry of school furniture, in the company Nautilus SA, located in Gondomar. She is currently working as a CAD Designer at Simoldes Plásticos, a company that manufactures components and accessories for motor vehicles.

Bárbara Rangel who is an architect is an Assistant Professor at the Civil Engineering Department of the Faculty of Engineering of the University of Porto (FEUP) since 2004, where she teaches courses about Architecture; Technical Drawing and Industrial Design in the Master of Civil Engineer, Master of Industrial and Product Design, and in the Ph.D. program with MIT Engineering Design and Advanced Manufacturing. She is member of the Executive Committee of Civil Engineering Department responsible for the communication and dissemination. She is an integrated member of CONSTRUCT (Institute of R&D in Structures and Construction). She is editor of Cadernos d'Obra, an International Scientific Journal, and responsible for the edition of scientific books: Livros d'Obra and Sebentas d'Obra. She is Guest Editor of Springer editions. She is coordinator of two European projects: Community Service Engineer (539642-LLP-1-2013-1-BE-ERASMUSEQR) and Educational Lab—Big Machine Erasmus+ project (2016-1-PT01-KA201-022986). Her main research interests are integrated project delivery, incremental housing, architectural, and project management, industrial design, and project-based learning. Faculdade de Engenharia da Universidade do Porto DEC, Rua Dr. Roberto Frias 400, 4200-465 Porto, Portugal.

Vitor Carneiro is a Product/Industrial Designer at NAUTILUS S.A., a Portuguese company of school furniture and educational technologies, since 2015, where he actively collaborates in the development and production of new and innovative products for schools. His passion for design started earlier and by accident, but quickly becomes a passion. He has been involved in several projects, designing and developing products that make a difference in someone's life, having been that effort recognized, for example, with Creativeness in Ergonomics Student of The Year Award (2015), along with two colleagues.

Jorge Lino Alves is a Researcher at INEGI/LAETA and has been an Associate Professor at the University of Porto (DEMec/FEUP/UPorto) since 2004, in Porto, Portugal, where he teaches courses about Materials and Industrial Design. He is Adjunct Director of the Master Program in Product and Industrial Design, Director of DESIGNSTUDIO FEUP, and Vice President of the Portuguese Society of Materials (SPM). He is an integrated member of INEGI/LAETA (Institute of Science and Innovation in Mechanical and Industrial Engineering/Associated Laboratory for Energy, Transports, and Aeronautics). His main research interests are additive manufacturing, industrial design, materials, new technological processes, and new methodologies in teaching engineering. INEGI, Faculdade de Engenharia da Universidade do Porto, Rua Dr. Roberto Frias 400, 4200-465 Porto, Portugal.

The Views of Engineering Students on Creativity

Paula Catarino, Maria M. Nascimento, Eva Morais,
Paulo Vasco, Helena Campos, Helena Silva, Rita Payan-Carreira
and M. João Monteiro

Abstract Creativity plays a growing role in education, from elementary school to higher education. Nowadays, both employers and universities develop research and are committed to the development of the twenty-first-century interpersonal, applied skills—creativity included—foreseen as fundamental to all professionals, engineers added. Generally, engineering degrees focus on the content of their scientific areas. In some higher education degrees, creativity still plays a small role. In order to reinforce the importance of creativity in the engineering degrees in a Portuguese northeastern university, it was pertinent to study the conceptions of engineering students about creativity. This study presents the conceptions of creativity of the first-year students of higher education, in the engineering area in two school years. The answers of 128 first-year students from two academic years (61 from 2014/15

P. Catarino (✉) · M. M. Nascimento · E. Morais · P. Vasco · H. Campos
H. Silva · R. Payan-Carreira · M. João Monteiro
Departamento de Matemática da Escola de Ciências e Tecnologia da, Vila Real Universidade
de Trás-os-Montes e Alto Douro—UTAD, Portugal Quinta de Prados, 5000-801 Vila Real,
Portugal
e-mail: pcatarin@utad.pt
URL: www.utad.pt

M. M. Nascimento
e-mail: mmsn@utad.pt

E. Morais
e-mail: emorais@utad.pt

P. Vasco
e-mail: pvasco@utad.pt

H. Campos
e-mail: hcampos@utad.pt

H. Silva
e-mail: helsilva@utad.pt

R. Payan-Carreira
e-mail: ritapay@utad.pt

M. João Monteiro
e-mail: mjmonteiro@utad.pt

© Springer Nature Singapore Pte Ltd. 2018
M. M. Nascimento et al. (eds.), *Contributions to Higher Engineering Education*,
https://doi.org/10.1007/978-981-10-8917-6_6

135

and 67 from 2016/17) and four different degrees to the open question—"What is creativity?" were analyzed. It was a mixed study, qualitative to deepen students' conceptions and quantitative to study some proportions differences and variables crossing. The results show low personal involvement even in the use of the first person plural in either school year, although the students' most used sentence was "for me." In both academic years, students' definitions mentioned more the creation of the implicit category in the content analysis. The words "new" and "way" were common to all the word clouds produced, and creativity and innovation appear somehow connected. In general, proportion differences were not statistically significant and degree crossed with categories showed no dependency.

Keywords Mathematical · Creativity · Engineering education · Students Conceptions

1 Introduction

Currently, it gets increasingly difficult for teachers to motivate students and encourage the learning of several technical subjects important for their future. The use of different methodologies and learning contexts are essential to assist the teaching and learning process and to promote the development of creativity on students and future engineers.

Creativity is an essential ingredient in modern societies, but its definition is not simple. Today, the importance of creativity has been acknowledged, particularly in an educational context. We should motivate our students for learning the different subjects, developing their creativity, and promoting their academic success. Simultaneously, it is important that higher education keeps the pace with the engineers' labor market, in order to prepare everyone, society and people, for new changes fostering innovation and creativity.

Some researchers are focused on developing a clear concept of the terms creativity and innovation and investigate their possible relation with engineering education (e.g., [1–9]). However, in order to help promoting creativity in engineering education, we believe that it is essential to know which are the conceptions of creativity in engineering students. Therefore, to address the conceptions of creativity, we developed a study centered in an open-question survey presented to students of the first year of higher education in the engineering area, in two consecutive academic years. Stepping from the results, we reflected on how to enhance creativity in engineering education from different approaches in different subjects of the curricula.

2   Theoretical Framework

According to Morell [10], "Engineering education (…) plays a central role in our increasingly technology-based societies. The education of engineers must prepare them for the multidisciplinary nature of the problems they will face developing a new set of skills and competencies." Morell [10] also

> lists five things engineering education can do to better respond to society's needs: innovate, reform the engineering curriculum and the learning experience, focus on learning (not on teaching), foster creativity and innovation across the ecosystem, implement continuous assessment and accreditation to drive excellence and educate the engineering professor of the future.

More recently Sola [11, p. 11] recognized that

> Engineers are in the business of innovation, and creativity is the foundation of that business. With this foundation, engineers create the solutions needed to address the challenges of the world. To better understand the implications of creativity and innovation, we must first understand what these concepts truly encompass.

Today, in any profession, creativity must be taken into account, because only creative people can be different, boost their careers, innovate in their jobs, creating new things and solving problems. Then, one may question what creativity is for the future engineer.

2.1   What Is Creativity?

It is difficult to find a consensual definition for "creativity." We understand that "Creativity is an essential ingredient of modern societies, associated with progress in the general welfare at the population level, since it may give answers to the present and future requirements." [12, p. 864].

Paul Torrance was a pioneer researcher on creativity, who dedicated "his life's work to study the nature of creativity and how it can be taught to students of all ages" ([12], p. 1). According to Torrance (1963, quoted by Stouffer et al. [13, p. 1]), creativity is "the process of sensing problems or gaps in information, forming ideas of hypotheses, testing, and modifying these hypotheses, and communicating the results. This process may lead to any one of many kinds of products—verbal and nonverbal, concrete and abstract."

Several other researchers gave definitions for creativity. For example, for Farid et al. [14], creativity is "the awareness, observation, imagination, conceptualization, and rearrangement of existing elements to generate new ideas." For Court [15, p. 141], "creativity may also be considered as a physical process that one must undertake to achieve a particular goal, as well as an individual quality that one naturally possesses." Several papers are devoted to creativity and different conceptions/definitions may be found (e.g., [3, 8, 16–18]). Klausen [18] attempted to define and understand creativity, informed by the methods and debates of contemporary philosophy. Sometimes, creativity is associated with the art and literature [19], but actually, creativity is also associated with the science area. Starko [20] defined creativity as the development of ideas that are novel and appropriate.

Some authors offered other definitions of creativity, such as the one proposed by Vernon (quoted by Lai [21]), where creativity is "a phenomenon related to the ability to produce ideas (imagination), restructuring (innovation), discoveries (inventions), new and original artistic objects (creation and originality), and all these types of abilities are needed for thinking (thinking, therefore critical thinking)."

Some studies were developed with students in all levels of education to know their definitions of creativity (e.g., [22]). Using Vernon's definition, Catarino et al. [23] studied the conceptions of creativity of university students' in the first year of engineering subjects. The results showed that the students' definitions "were affected neither by gender nor by the original area of study and both genders and undergraduate course showed the predominance (mode) of grouped implicit categories (creation, imagination, and originality)."

In the current study, we extended Catarino et al. [23] work, adopting once more the Vernon's definition, for creativity.

2.2 Is Creativity Important in the Performance of an Engineer?

Cropley et al. [1, p. 211] state

> At the level of the individual engineer, considerations of the global marketplace and the creative skills regarded as essential for a successful career in engineering have also raised the issue of fostering creativity in engineering education.

Cropley et al. [1, p. 210] defend that "the most important characteristic of engineering creativity is to perform tasks or solve problems. In solving problems, any new and useful ideas could be potential solutions. The value of creativity could be helpful for solving problems, whereas one could also accept problem-solving as one kind of creativity."

According to Maiden et al. (2001 quoted by Dallman et al. [24]), creativity is viewed as a central part of requirements engineering (RE) and call for the recognition of the creativity importance on the RE process. Dallman et al. [24] revised the understandings of creativity in RE and suggested that further understanding of the nature and context of creativity is required to promote and encourage creative RE practices. The work of Nguyen and Shanks [25] also built a theoretical framework for understanding creativity in RE that included five elements: product, process, domain, people, and context. They defend that "RE researchers and practitioners need to recognize different creativity elements and integrate them within RE approaches. Creativity elements should be applied at an appropriate situated level with a view to developing ICT[1] innovations for business" [25].

Some of the questions raised by Stouffer et al. [13] in 2004 remain unanswered today. One of them was "what is creativity and how can you teach it to engineering students?" Stouffer et al. [13, p. 1] examined "these questions to make the case that fostering creativity knowledge, skills, and attitudes is vital for the future of engineering and engineering education."

Systematic creative methods do exist to fill the lack of creativity in graduating engineers; one of them is the theory of inventive problem-solving (TRIZ), introduced by Ogot and Okudan [26] in the first year of a student's academic career in a required Introduction to Engineering Design course. This course "employs a design-driven curriculum with emphasis placed on skills such as teamwork, communication skills (graphical, oral, and written), computer-aided design and analysis tools" [26, p. 109]. The results showed that the first-year students adhered in a positive way to TRIZ as a creativity method.

The work of Badran [3] focused on developing a clear concept of the terms creativity and innovation, and investigate their possible relation with engineering education. He concludes his study stating that engineering education should introduce "relevant co-curricular multidisciplinary activities, engineering projects at all levels, early exposure to industry."

Other researchers studied the competencies required by engineers. For example, Male et al. [27] presented an Australian study on the generic competencies required by engineers. Their results indicated that both the non-technical, attitudinal and the technical competencies were perceived as important. Some examples of attitudes are commitment, honesty, self-motivation, demeanor, creativity, and concern for others. Also, two of the major competency factors identified as important to the work of engineers were "Creativity/Problem-solving" and "Innovation." In this study, the generic graduate attributes corresponding to "Creativity/Problem-solving" were "Ability to undertake problem identification, formulation, and solution; ability to utilize a systems approach to design; and operational performance" [27, p. 160].

Zhou [4] focused on the question of how engineering students perceive the strategy of integrating creativity training into a problem- and project-based learning curriculum. Results showed that the training program was thought useful and students got benefits such as gaining project work skills, creative concepts, and confidence in being creative.

Ibrahim [28], who explored the relationships among creativity, engineering knowledge, and team interaction on senior engineering design product outcomes, stated that "A better understanding on the interaction of these three constructs would help engineering educators to design and establish a better curriculum for our future engineering student candidates" [28, p. 180].

---

[1]Information and communications technology.

Fostering creativity in the engineering classroom had also lead to more successful students and better student-professor interactions. For Cropley [8, p. 155], "[c]reativity is a fundamental element of engineering", and "of special importance is embedding creativity in engineering education". In Cropley's [8, p. 160] opinion, "two basic components are needed by engineers entering the field of creativity to answer the question what is creativity?" The first component was clearly described by Plucker et al. [29]—"[c]reativity is the interaction among aptitude, process and environment by which an individual or group produces a perceptible product that is both novel and useful as defined within a social context" (p. 90)—while the second component is characterized by the 4Ps: "Person, Product, Process and Press (environment)" [8, p. 161].

For Charyton [30, p. 135], "a creative act needs acceptance of an idea, product, or process by the field, such as engineering, and the domain, such as science or Science, Technology, Engineering and Mathematics (STEM). Today's engineers must be creative and innovative. The problems faced by engineers today request original thinking. To remain competitive globally, engineering firms rely on creative individuals and creative teams to develop new products for innovation."

According to Cropley [8, p. 161] "creativity in engineering is concerned with solving problems; however, the solutions engineers devise do not emerge in a single step. Engineers understand that there is a sequence of stages that is followed starting with the recognition that there is a problem to be solved, and followed by the determination of possible ways of solving that problem, narrowing these down to one, or a few, probable solutions, before selecting the best option for development and implementation. Creativity in engineering is embedded across this sequence of stages"

The "problem-solving" is an important technique that should be implemented in all subjects within any engineering curricula. Karataş et al. [31] on the views of the nature of engineering and their implications for engineering degrees noticed that the students in their study held tacit beliefs that engineering is a form of applied science, but our emphasis goes what they stated as it involves problem-solving and design of artefacts or systems subject, and also referred that teamwork is needed.

Recently, the empirical study presented by Martín-Erro et al. [32], with engineering students and professors, rejects the first idea reported in the literature that "creativity is not valued into engineering educational environment, which shows an evolution on engineering student's opinions," That study revealed efforts made by professors in teaching according to creative approaches, such as the project-based learning and open problems, which confirm their interest on enhancing creativity. The majority of the participants (95%)—professors and students—agreed that creativity is important for engineering; 92% of the professors also agreed that is more important to stimulate creativity through the practice than by teaching creative techniques.

In the opinion of Sola [11] "[c]reativity and critical thinking are essential tools for engineers. Without them, engineers may face challenges that they are ill-prepared to solve. By better understanding the various aspects of creativity and critical thinking, (…) engineers can improve their problem-solving performance.

This is not only beneficial for these individuals, but is bound to also provide benefits to everyone through their resulting ground-breaking discoveries and solutions" [32, p. 127].

In summary, we may say that both employers and universities have been researching and committed with the development of the twenty-first-century interpersonal, applied skills—creativity included—required to all professionals, but specifically to engineers. As written by Kivunja [33], "the 4Cs of critical thinking and problem solving, communication, collaboration, and creativity plus innovation, [are] the super skills in the 21st century because they are foundational essentials for success in college, university, career, and life outside educational institutions."

In line with this framework, this study focused on the conceptions of creativity held by first-year students of higher education collected during two academic years (2014/2015 and 2016/2017), considering different genders and undergraduate courses. The creativity remains unquantifiable (e.g., [34, 35]). With this aim, we decide to do an exploratory study based on these first-year undergraduates' conceptions about "creativity," following the Vernon categories (Vernon, 1989, quoted by Lai [21]).

3   Research Methodology

3.1   Research Participants

An online survey on Google Drive (GD) was made available to a group of first-year engineering students from a northeastern Portuguese university that enrolled a Linear Algebra (LA) course. In 2014/2015, a total of 61 students from four degrees participated in the survey (Table 1): Biomedical Engineering (Biomedical Eng., 27.9%; $n = 17$); Bioengineering (Bioeng., 19.7%; $n = 12$); Mechanical Engineering (Mechanical Eng., 37.7%; $n = 23$); and Energy Engineering (Energy Eng., 14.8%; $n = 9$). Men were the mode (41 men, 67.0%, vs. 20 women, 33.0%). Men were the mode (41 men, 67.2%, and 20 women, 32.8%). The participants' ages ranged between 17 and 25 years old, although women were younger, ages ranged between 17 and 19 years old.

In 2016/2017, a total of 67 students from four degrees participated in the survey (Table 1): Biomedical Eng. (26.9%); Bioeng. (20.9%); Mechanical Eng., 27 students, 40.3%; one woman, 3.7%, 96.3% men); and Integrated Master's in Industrial Management and Engineering (MIEGI, 11.9%). Men were also the mode (29 women, 43.3% and 38 men, 56.7%). The age of participants ranged between 17 and 26 years old; also, in this academic year, the women were younger, presenting the same age range as in the previous year.

Table 1 A simple overview of the dimension and hierarchy levels related to engineering schools

School year	Degree	Female (%)	Male (%)	Total (%)
2014/2015	Biomedical Eng.	58	42	17 (27.9)
	Bioeng.	67	33	12 (19.7)
	Mechanical Eng.	4	96	23 (37.7)
	Energy Eng.	11	89	9 (14.8)
2016/2017	Biomedical Eng.	88.9	11.1	18 (26.9)
	Bioeng.	64.3	35.7	14 (20.9)
	Mechanical Eng.	3.7	96.3	27 (40.3)
	MIEGI	37.5	62.5	8 (11.9)

3.2 Data Collection and Analysis' Steps

The survey's participants in either academic year shared the same LA teacher; some students had two or three registrations in the LA course. To compare the results with data issue from Catarino et al. [23] study, the same methodology of content analysis was followed. The content analysis disclosed groups of words in the students' texts respecting their understanding of creativity; thereby, the categories derived from the raw data in each student text as well as creativity definitions [23, 36]. The survey asked the students to answer by their own words to the question "What do you understand by creativity?" besides recording also their gender and their age. The questionnaire had to be answered in the last 15 min of a LA class, in the first semester of the academic years 2014/2015 and 2016/2017. It is important to say that no creativity definitions were given to the students and they had to write it individually.

A mixed study was implemented, qualitative to deepen students' conceptions on creativity and quantitative to study the differences between some proportions and variables crossing [36]. As qualitative methods, content analysis and words clouds were used. The students' answers (128 in total, 61 in 2014/2015, and 67 in 2016/2017) were categorized in order to describe their understanding of creativity. Two of the authors did the categorization for each answer inductively based on all of the written text: words and sentences. After that, the other authors confirmed the categorization and the subcategories were established. Each answer was included in a subcategory. Finally, due to the amount of subcategories, the authors agreed to reduce them to broader categories. Therefore, this content analysis was applied "to the manifested content that is to words, paragraphs, and sentences written, and we established the content analysis categories" [23, 37], which was students' written understanding of creativity. We read all the 128 texts word-by-word and derived codes by highlighting their important meanings, and all of them were considered anonymously. Next, we analyzed the personal involvement of the students, as in Maksić and Pavlović [38]. Finally, we adopted the same categories for creativity that emerged in Catarino et al. [23], namely "implicit" (creation, imagination, and originality) and "explicit" (innovation, inventions, and thinking). The content analyses were performed directly in a spreadsheet, since it was the original format downloaded from GD. Schematically, the students' definitions are given using a word cloud, where "words are arranged artistically in close proximity and the size of each word's type is proportional to the word's frequency or to the size of a numeric variable associated with the word" [39].

The quantitative analysis used counting, percentages, tables, graphs, and crossed tables in IBM-SPSS version 24. Collected data were tested for differences between the categories proportions (p value < 0.05, the proportions were considered different) for either the total sample (all the 128 students) or the academic year (61 students in 2014/205 and 67 students in 2017/2016). Each subset of variable versus categories was tested for independence (p value < 0.05, the variables were dependent).

4   Data Analysis and Discussion

Like in Maksić et al. [38], the study began with the analysis of the existence and categorization of the personal involvement established in definitions of creativity of the students. Next, the implicit and explicit Vernon categories [21] were analyzed crossing the results with gender and degrees, and finally, we build and analyze the words clouds. In each section, we present tables with examples of categories options in Portuguese followed by the translation into English.

4.1   Personal Involvement in the Creativity Definitions

Most students answered the survey with sentences or paragraphs. Some words and sentences used by the students were such as "for me "I believe that," "in my opinion," "in my view," "from my perspective," "I understand that," "I consider that" (Tables 2 and 3). Involving us all, we also considered the use of the first person plural (Table 3). The text of the students also may be considered a writing style artifact, but they reflect the view of the student and their way of writing about it. In order to explain the categories in this section, a single example of the expressions implying the students' personal involvement in their creativity definitions is presented both in English (translation) as in Portuguese (original sentence, Tables 2 and 3).

In Table 4, we listed the students' definitions denoting personal involvement, which were used by 17 of 61 students (28%) in 2014/2015 and 24 of 67 students (36%) in 2016/2017. The use of the first person plural was another group considered: 4 of 61 students (7%) in 2014/2015, and 3 of 67 students (4%) in 2016/2017. In both academic years, 40 students provided non-personal definitions, representing 65% of the students in 2014/2015 (40 of 61 students) and 60% in 2016/2017 (40 of 67 students); but no differences were found between the two years ($p = 0.56$). In the

Table 2 Words and sentences implying personal involvement of the students

Words and sentences	2014/2015		2016/2017	
	Example of participants' answers	Original sentence in Portuguese	Example of participants' answers	Original sentence in Portuguese
"for me"	To me, creativity is a person's ability to have new ideas, different, interesting, to have the power to change the world and to be useful to us in our day-to-day	Para mim a criatividade é a capacidade de uma pessoa ter ideias novas, diferentes, interessantes, capaz de poder revolucionar o mundo e de ser-nos útil no nosso dia-a-dia	To me, creativity is to be able to create new things, to imagine beyond the possible and to try to make it come true. A way to face everyday life in a different way, a way to create change, one of the ways to see beauty in smaller things	Para mim a criatividade é ser capaz de criar coisas novas, imaginar para lá do possível e tentar torná-lo realidade. Uma maneira de olhar para o quotidiano de outra forma, uma maneira de criar a mudança, uma das maneiras de ver a beleza nas coisas mais pequenas
"I believe that"	I believe that creativity grows with the experience and practice of new ideas	Acredito que a criatividade cresce com a experiência e com a prática de novas ideias	–	–
"in my opinion"	In my opinion, creativity is when, thinking at any theme, a number of ideas arise about how to do something fun and complete, full of imagination!	Na minha opinião a criatividade é quando, ao pensar num tema qualquer, nos surgem diversas ideias sobre como fazer algo divertido e completo, cheio de imaginação!	In my opinion, creativity is a tool that fewer and fewer people have. Another way to look at all-day life, a way to create change, one of the ways to see beauty in smaller things	Na minha opinião, criatividade é uma ferramenta, que cada vez mais menos pessoas a possuem. Uma maneira de olhar para o quotidiano de outra forma, uma maneira de criar a mudança, uma das maneiras de ver a beleza nas coisas mais pequenas
"in my view"	In my view, creativity is to achieve what has not been achieved by any other person	A meu ver, criatividade consiste em alcançar o que ainda não foi alcançado por qualquer outra pessoa	In my view, creativity is the art of imagining far away, is having the ability to create something new…	No meu ponto de vista a criatividade é arte de imaginar mais além, é ter a capacidade de criar algo novo…

Table 3 Words and sentences implying personal involvement of the students (conclusion)

Words and sentences	2014/2015		2016/2017	
	Example of participants' answers	Original sentence in Portuguese	Example of participants' answers	Original sentence in Portuguese
"from my perspective"	Creativity, from my perspective, is something that allows us to innovate, something that makes us different from others, because when we are creative, we do things different from others	A criatividade, na minha perspetiva, é algo que nos permite inovar, algo que nos torna diferente dos outros, porque ao sermos criativos, fazemos coisas diferentes dos outros	–	–
"I understand that"	I understand that creativity is to be original and think differently from others. Think by his own head and be able to see that nobody saw	Entendo que criatividade é ser original e pensar de forma diferente dos outros. Pensar pela sua própria cabeça e ser capaz de ver o que ainda ninguém viu	–	–
"I consider that"	I also consider that creativity exists in all the people and in all the situations, because when several people have something to do, none of them do those things the same way as others, so we are all creative, and thanks to this creativity is we are to evolve scientifically, because creativity is the greatest tool of science, art, is "out for adventure"	Considero também, que a criatividade existe em todas as pessoas e em todas as situações, visto que quando várias pessoas têm algo para fazer, nenhuma destas pessoas faz essas coisas igual as outras, por isso todos nós somos criativos, e graças a essa criatividade é que estamos a evoluir cientificamente, pois a criatividade é a maior ferramenta da ciência, da arte, é o "sair para a aventura"	–	–

(continued)

Table 3 (continued)

Words and sentences	2014/2015		2016/2017	
	Example of participants' answers	Original sentence in Portuguese	Example of participants' answers	Original sentence in Portuguese
Use of the first person plural	Creativity is our ability to innovate and create new ideas and/or products	Criatividade é a capacidade que possuímos de inovar e criar ideias e/ou produtos novos	Creativity is the way we apply our skills in a creative process, the ease with which we solve or create something instantly and well	Criatividade é a forma como aplicamos as nossas aptidões num processo criativo, a facilidade com que se resolve ou se cria algo de forma instantânea e bem conseguida

Table 4 Words and sentences implying personal involvement of the students in their conceptions of creativity

Words and sentences	Number of references (%)	
	2014/2015	2016/2017
"for me"	8 (38.0)	21 (77.0)
"I believe that"	1 (4.8)	–
"in my opinion"	3 (14.3)	2 (8.0)
"in my view"	1 (4.8)	1 (4.0)
"from my perspective"	1 (4.8)	–
"I understand that"	2 (9.5)	–
"I consider that"	1 (4.8)	–
Use of the first person plural	4 (19.0)	3 (11.0)
Total	21 (100)	27 (100)

category of personal involvement, the sentence most used by students was "for me" with a bigger weight in 2016/2017 (77% vs. 38% in 2014/2015).

As in Maksić and Pavlović [38], we also observed that students frequently use relatively low levels of personal involvement in their creativity concepts. Nevertheless, in 2016/2017, the percentage of students using expressions with personal involvement was higher than in 2014/2015 ($p < 0.001$).

4.2 Outlook of Students' Creativity Definitions

A single example retrieved from the student's sentences following the Vernon's creativity definition is presented in Tables 5 and 6 for each of the considered categories, in both the English translation and the Portuguese original words or sentences.

Table 5 Words and sentences with examples of the students' definitions about creativity implicit categories by school year

Groups	Categories	2014/2015		2016/2017	
		Example of part of an answer	Original sentence in Portuguese	Example of part of an answer	Original sentence in Portuguese
Implicit	Creation	Creativity is the ability to create	Criatividade é a capacidade de criar	Creativity is the ability to accomplish something in several possible and different ways	Criatividade é a capacidade de realizar algo de várias maneiras possíveis e diferentes
	Imagination	It is also a "way" to practice our imagination	É ainda uma "forma" de pôr em prática a nossa imaginação	Creativity is the ability to imagine something new, or existing in its own way. To be able to pick up an object and change it so that it has something of itself	A criatividade é a capacidade de imaginar algo novo, ou existente à sua maneira. Poder pegar num objeto e alterá-lo, de forma a ficar com algo de si mesmo
	Originality	To be original	Ser original	Creativity is an original way of facing obstacles and overcoming them in a different way. It is an unusual form of thinking and a way of being active… because nowadays we have to excel by originality	Criatividade é uma forma original de encarar os obstáculos e ultrapassa-los de forma diferente. É uma forma invulgar de pensar e uma maneira de estarmos ativos na sociedade em que vivemos, pois hoje em dia temos de primar pela originalidade

Table 7 presents the counting for the categories and subcategories contained on Vernon creativity definition [21]. In each academic year, the mode in the implicit categories was creation (2014/2015: 27%; 2016/2017: 29%). In addition, in both

Table 6 Words and sentences with examples of the students' definitions about creativity explicit and others categories by school year

Groups	Categories	2014/2015		2016/2017	
		Example of part of an answer	Original sentence in Portuguese	Example of part of an answer	Original sentence in Portuguese
Explicit	Innovation	Creativity or to be creative, is the ability to innovate, to do something that does not exist or change a reality and give it a new use	Criatividade ou ser criativo, é a capacidade de inovar, fazer algo que não existe ou alterar uma realidade e dar-lhe uma nova utilidade	Creativity is the ability to create new things, to innovate, to go further, to think of something that no one had ever thought [before]	A criatividade é a capacidade criar coisas novas, de inovar, de chegar mais longe, pensar em algo que nunca ninguém tinha pensado
	Inventions	Creativity is the creation of a new and different solution/idea to solve a given problem, that is, "invent" something that can effectively solve our obstacle	Criatividade é a criação de uma nova e diferente solução/ideia para resolver determinado problema, isto é, "inventar" algo que possa eficazmente solucionar o nosso obstáculo	Ability to invent something new, using the imagination	Capacidade de inventar algo novo, recorrendo à imaginação
	Thinking	Creativity is the ability that one has to put what it is, feels, thinks and argues in everything we do, so that the end result also shows what the person is	A criatividade é a capacidade que cada um tem de pôr aquilo que é, sente, pensa e defende em tudo aquilo que faz, de forma a que o resultado final mostre, também, aquilo que a pessoa é	Creativity to me is the inventiveness that we all have inside our head, from which we can develop ideas that can become something important or not for society	Criatividade, para mim, é um engenho que temos todos dentro da cabeça, a partir do qual podemos desenvolver ideias que se podem tornar algo importante ou não para a sociedade
Others		Creativity is almost like a gift	A criatividade é quase como um dom	It is the ability to wander in a world with a variable dimension depending on the rational understanding of each one	É a capacidade de divagar num mundo com dimensão variável dependendo da compreensão racional de cada um

Table 7 Words and sentences according to Vernon's definition of creativity

Groups	Categories	Number of references					
		2014/2015			2016/2017		
		Female	Male	Totals (%)	Female	Male	Totals (%)
Implicit	Creation	10	21	31 (27)	8	12	20 (29.0)
	Imagination	7	6	13 (11)	3	2	5 (7.2)
	Originality	9	15	24 (21)	1	6	7 (10.1)
Explicit	Innovation	8	14	22 (19)	11	9	20 (29.0)
	Inventions	6	5	11 (10)	3	2	5 (7.2)
	Thinking	1	1	2 (1)	3	7	10 (14.5)
Others		1	12	13 (11)	0	2	2 (3.0)
Total (%)		42 (36)	74 (64)	116 (100)	29 (42)	40 (58)	69 (100)

academic years in the explicit categories, the mode was innovation (2014/2015: 19%; 2016/2017: 29%). However, in neither case, the proportion differences had statistical significance ($p = 0.56$ and $p = 0.13$ for the implicit and explicit categories, respectively). The percentages of the categories by gender are presented in Table 7.

The students' definitions of creativity were independent from gender in either academic year (2014/2015: chi-squared = 0.21, $p = 0.65$; 2016/2017: chi-squared = 0.86, p = 0.36).

The implicit grouped categories were predominant (mode) in both genders in 2014/2015, Fig. 1 (left). Curiously, in 2016/2017, the mode was the implicit grouped categories for men, while for female, the mode was the explicit grouped categories, Fig. 1 (right).

Even though devoid of statistical significance, differences were observed in the way that men and women value some attributes within creativity categories. Within the implicit categories, male undergraduates valued less the imagination attributes than female counterparts (2014/2015: 8.1% vs. 16.7%, $p = 0.05$; 2016/2017: 5% vs. 10%, $p = 0.45$). Within the explicit categories, the term inventions was used less

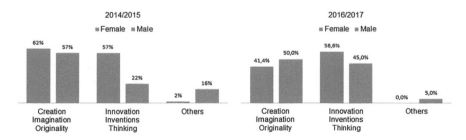

Fig. 1 Grouped categories by gender and by academic year

frequently by males than females (2014/2015: 6.8% vs. 14.3%, $p = 0.056$; 2016/2017: 5% vs. 10%, $p = 0.45$). Interestingly, statistical differences were found in the use of terms not included on Vernon's based categories—named as "others"—which was more frequent in male undergraduates' students than in female's, but only in 2014/2015 (2014/2015: 16.2% vs. 2.4%, $p = 0.011$; 2016/2017: 5% vs. 0%, $p = 0.148$).

For the word clouds, the same methodology as for the counting and graphs was used: by gender and by degree. Figure 2 presents the word cloud representing the female students' definitions according to the academic year. In 2014/2015, the word "way" stood out followed by "ideas," "imagination," and "original" while with a lower emphasis appeared words like "different" and "thinking." In 2016/2017, the word "new" stood out, followed by "different," "things," "way" and "ideas" and "innovate." Despite differently sized and colored in the images, "way" and "new" were common in both school years.

In line with several authors (e.g., [3, 11, 30, 33]), the female students' definitions of creativity mentioned innovation (or a similar word); its inclusion was more frequent in 2016/2017 than in 2014/15.

Figure 3 presents the word cloud for the male students' definitions of creativity by academic year. The word "new" stands out in both images followed by "way," "ideas" and "original," and finally "person" in both years; with a lower emphasis appear the words "think," "imagination," and "innovative." Curiously, despite the differences in the colors and location within the image, the word clouds for male students highlights the same words, and as occurring in females (Fig. 2), "way" and "new" are common to both academic years.

Once again, as stressed by several authors (e.g., [3, 11, 30, 33]), these students' definitions of creativity mentioned the term innovation (or a similar word).

Fig. 2 Word clouds for female students by academic year

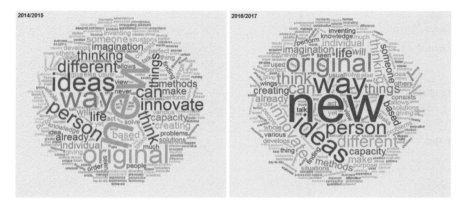

Fig. 3 Word clouds for male students by school year

Table 8 summarizes the results for the creativity definitions according to the students' degree. The two categories (implicit and explicit) were independent from the students' courses (chi-squared = 1.8, p = 0.61, 2014/2015; chi-squared = 0.37, p = 0.54, 2016/2017).

Figure 4 shows the predominance of the implicit grouped categories in all the degrees in 2014/2015 and in 20167207, except for the few students of MIEGI that favored the explicit grouped categories. Distribution in the Bioengineering students' definitions about creativity was uniform among the three groups analyzed in 2014/2015, but in 2016/2017 with fewer students, the distribution is irregular. Those students referred to imagination more often in 2014/2015 (18.4%), but did not refer it at all in 2016/2017. Regarding invention, 13.1% of the Bioengineering students mentioned it more frequently than in the other undergraduates' courses in 2014/2015, while in 2016/2017, this same category had very few references. Only the Biomedical Engineering students mentioned the thinking category in 2014/2015; in 2016/2017, only MIEGI students failed to refer it.

The word clouds were only built for the groups of Bioengineering and Biomedical Engineering together in both academic years and for Mechanical Engineering in both school years.

The word clouds issued from the definitions of creativity by the Bioengineering and Biomedical Engineering students taken together by academic year are presented in Fig. 5. In both years, the word "new" popped up, in 2014/2015 followed by "different" and in 2016/2017 by "ideas." Next, the word "way" appeared followed by different words in 2014/2015 "things," "ideas," and "unique" and in 2016/2017 "innovative," "imagination," and "original" and with a lower emphasis appear the words "different" and "thinking." Despite being differently colored, "way" and "new" are common to the definitions in both academic years, supporting the mentioned by some (e.g., [3, 11, 30, 33]) respecting the close association of innovation (or a similar word) to creativity in definitions, particularly for the texts collected in 2016/2017.

Table 8 Global results for the students' definitions of creativity according to their degree and academic year

Groups	Categories		2014/2015		
		Bioeng. [Biomedical Eng.]	Mechanical Eng.	Energy Eng.	Totals (%)
Implicit	Creation	8 [4]	12	7	31 (27)
	Imagination	7 [2]	3	1	13 (11)
	Originality	7 [5]	9	3	24 (21)
Explicit	Innovation	7 [6]	6	3	22 (19)
	Inventions	5 [6]	3	2	11 (9)
	Thinking	0 [6]	0	0	2 (2)
Others		4 [6]	7	0	13 (11)
Totals (%)		38 (33) [22 (19)]	40 (34)	16 (14)	116(100)
Groups	Categories		2016/2017		
		Bioeng. [Biomedical Eng.]	Mechanical Eng. [Energy Eng.]	MIEGI	Totals (%)
Implicit	Creation	5 [5]	7	3	20 (29)
	Imagination	0 [3]	2	0	5 (7)
	Originality	3 [1]	3	0	7 (10)
Explicit	Innovation	2 [6]	10	2	20 (29)
	Inventions	1 [2]	0	2	5 (7)
	Thinking	4 [1]	5	0	10 (15)
Others			1 [0]	0	2 (3)
Totals (%)			16 (23) [18 (26)]	27 (39)	69(100)

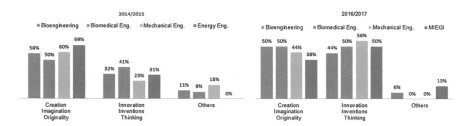

Fig. 4 Grouped categories by degree and academic year

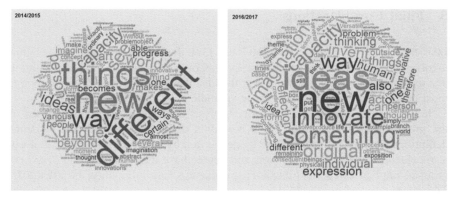

Fig. 5 Word clouds for Bioengineering and Biomedical Engineering students by academic year

The word clouds for the Mechanical Engineering students by school year are sketched in Fig. 6. In both school years, the word "new" also popped up, in 2014/2015, followed by "way" and in 2016/2017 by "things." In both academic years, the word "way" emerged (bigger in 2014/2015) and different words follow it. In 2014/2015, "person," "ideas," and "original, and in 2016/2017, "person," "imagine," and "capacity," and with a lower emphasis the words "different" and "thinking" appeared. Once again, despite the size and the colors, "way" and "new" are common in both school years.

Fig. 6 Word clouds for Mechanical Engineering students by school year

5   Final Considerations

Like the report by Maksić and Pavlović [38], few students used the personal involvement in their definitions of creativity. Conversely, we also found definitions that used the first person plural, giving the idea that creativity is a skill common to most people, but these were also more sporadic.

In the present study, the exploratory analysis of students' definitions, based on the grouped categories proposed in Vernon's definition, showed that students' definitions were not affected neither by gender nor by the original study field; it was also evidenced that both the genders and degree have the predominance (mode) of grouped implicit categories (creation, imagination, and originality) in 2014/2015. The same occurred in 2016/2017; an exception was the few MIEGI students who showed predominance (mode) of grouped explicit categories (innovation, inventions, and thinking).

The use of word clouds highlighted the focus put by students on words like "new" and "way" in their definitions of creativity. The tendency to connect creativity with the innovation (or a similar word) mentioned by some authors (e.g., [3, 11, 30, 33]) was further evidenced in the word clouds, namely on those representing the gender and the Bioengineering and Biomedical Engineering degrees. Other words less weighted—such as "thinking," "ideas," "imagination," "unique"—were mentioned in the students' definitions, as it was also referred by some authors (e.g., [14, 21]) in their texts about creativity.

Although this was an exploratory mixed study, it was interesting to note that in the students' definitions of creativity some of the literature mentions also arose.

The crossing of variables failed to evidence the existence of dependencies between variables, which suggests that either none existed because the students answered an open question without previous talking about creativity in classes or that the sampling of definitions analyzed was still in small number. So, further studies using the same methodology should be continued in future.

Starting with the students' definitions, we should be able to engage students in learning strategies or activities that may be perceived by them as empowering or encouraging creativity skills. Since in the profession, engineers' creativity is often challenged when facing problems, the use of problem-solving or project-based learning (e.g., [8, 31]) would foster students' confidence by training their openness to recognize different level of problems and by triggering their curiosity for new approaches toward common problems, and challenging their ability to propose and select the most suitable solution. We should work with our students as engineers!

Acknowledgements This work is financially supported by National Funds through FCT—Fundação para a Ciência e a Tecnologia under the project UID/CED/00194/2013 and PEst-UID/MAT/00013/2013.

References

1. Cropley DH, Cropley AJ (2000) Fostering creativity in engineering undergraduates. High Ability Stud 11(2):207–219. https://doi.org/10.1080/13598130020001223
2. Christiaans H, Venselaar K (2005) Creativity in design engineering and the role of knowledge: modelling the expert. Int J Technol Des Educ 15(3):217–236. https://doi.org/10.1007/s10798-004-1904-4
3. Badran I (2007) Enhancing creativity and innovation in engineering education. Eur J Eng Educ 32(5):573–585. https://doi.org/10.1080/03043790701433061
4. Zhou C (2012) Integrating creativity training into problem and project-based learning curriculum in engineering education. Eur J Eng Educ 37(5):488–499. https://doi.org/10.1080/03043797.2012.714357
5. Zhou C (2012) Fostering creative engineers: a key to face the complexity of engineering practice. Eur J Eng Educ 37(4):343–353. https://doi.org/10.1080/03043797.2012.691872
6. Walsh E, Anders K, Hancock S, Elvidge L (2013) Reclaiming creativity in the era of impact: exploring ideas about creative research in science and engineering. Stud High Educ 38 (9):1259–1273. https://doi.org/10.1080/03075079.2011.620091
7. Cropley DH (2015) Promoting creativity and innovation in engineering education. Psychol Aesthet Creat Arts 9(2):161–171. https://doi.org/10.1037/aca0000008
8. Cropley DH (2015) Creativity in engineering. In: Corazza GE, Agnoli S (eds) Multidisciplinary contributions to the science of creative thinking. Springer, London, pp 155–173
9. Kazerounian K, Foley S (2007) Barriers to creativity in engineering education: a study of instructors and students perceptions. J Mech Des 129(7):761–768. https://doi.org/10.1115/1.2739569
10. Morell L (2010) Engineering education in the 21st century: roles, opportunities and challenges. Int J Technol Engineering Educ 7(2):1
11. Sola E (2016) An experimental investigation of the state of creativity, critical thinking and creativity training in undergraduate engineering students. Dissertation, University of Central Florida
12. Cardoso AP, Malheiro R, Rodrigues P, Felizardo S, Lopes A (2015) Assessment and creativity stimulus in school context. Procedia-Soc Behav Sci 171:864–873. https://doi.org/10.1016/j.sbspro.2015.01.202
13. Stouffer WB, Russell JS, Oliva MG (2004) Making the strange familiar: creativity and the future of engineering education. In: Proceedings of the 2004 American Society for Engineering Education Annual Conference & Exposition, Salt Lake City, USA, 20–23 June 2004
14. Farid F, El-Sharkawy AR, Austin LK (1993) Managing for creativity and innovation in A/E/C organizations. J Manag Eng 9(4):399–409. https://doi.org/10.1061/(ASCE)9742-597X(1993)9:4(399)
15. Court AW (1998) Improving creativity in engineering design education. Eur J Eng Educ 23 (2):141–154. https://doi.org/10.1080/03043799808923493
16. Morrison A, Johnston B (2003) Personal creativity for entrepreneurship: teaching and learning strategies. Act Learn High Educ 4(2):145–158. https://doi.org/10.1177/1469787403004200
17. Runco MA, Jaeger GJ (2012) The standard definition of creativity. Creat Res J 24(1):92–96. https://doi.org/10.1080/10400419.2012.650092
18. Klausen SH (2010) The notion of creativity revisited: a philosophical perspective on creativity research. Creat Res J 22(4):347–360. https://doi.org/10.1080/10400419.2010.523390

19. Nadjafikhah M, Yaftian N, Bakhshalizadeh S (2012) Mathematical creativity: some definitions and characteristics. Procedia-Soc Behav Sci 31:285–291. https://doi.org/10.1016/j.sbspro.2011.12.056
20. Starko AJ (2005) Creativity in the classroom: schools of curious delight. Routledge, New
21. Lai ER (2011) Critical thinking: a literature review. Pearson's Res R 6. http://images.pearsonassessments.com/images/tmrs/CriticalThinkingReviewFINAL.pdf Accessed 15 Jun 2016
22. Gilat T, Amit M (2013) Exploring young students creativity: the effect of model eliciting activities, PNA 8(2): 51–59. http://hdl.handle.net/10481/29578
23. Catarino P, Nascimento MM, Morais E, Silva H, Payan-Carreira R (2016) Take this waltz on creativity: the engineering students' conceptions. In: Proceedings of the 2nd international conference of the Portuguese Society for engineering, IEEE Conference Publications, Vila Real 20–21 Oct 2016
24. Dallman S, Nguyen L, Lamp J, Cybulski J (2005) Contextual factors which influence creativity in requirements engineering. In: Proceedings of the 13th European conference on information systems, information systems in a rapidly changing economy, Regensburg, Germany, 26–28 May 2005
25. Nguyen L, Shanks G (2009) A framework for understanding creativity in requirements engineering. Inf Softw Technol 51(3):655–662. https://doi.org/10.1016/j.infsof.2008.09.002
26. Ogot M, Okudan GE (2006) Integrating systematic creativity into first-year engineering design curriculum. Int J Eng Educ 22(1):109–115
27. Male SA, Bush MB, Chapman ES (2011) An Australian study of generic competencies required by engineers. Eur J Eng Educ 36(2):151–163. https://doi.org/10.1080/03043797.2011.569703
28. Ibrahim B (2012) Exploring the relationships among creativity, engineering knowledge, and design team interaction on senior engineering design projects. Dissertation, Colorado State University
29. Plucker JA, Beghetto RA, Dow GT (2004) Why isn't creativity more important to educational psychologists? Potentials, pitfalls, and future directions in creativity research. Educ Psychol 39(2):83–96. https://doi.org/10.1207/s15326985ep3902_1
30. Charyton C (2015) Creative engineering design: the meaning of creativity and innovation in engineering. In: Charyton C (ed) Creativity and innovation among science and art. Springer, London, pp 135–152
31. Karataş FÖ, Bodner GM, Unal S (2016) First-year engineering students' views of the nature of engineering: implications for engineering programmes. Eur J Eng Educ 41(1):1–22. https://doi.org/10.1080/03043797.2014.1001821
32. Martín-Erro MA, Espinosa MM, Domínguez M (2016) Creativity and engineering education: a survey of approaches and current state. In: Proceedings of the 9th International Conference of Education, Research and Innovation, Seville, Spain 14–16 Nov 2016
33. Kivunja C (2015) Exploring the pedagogical meaning and implications of the 4Cs "super skills" for the 21st century through Bruner's 5E lenses of knowledge construction to improve pedagogies of the new learning paradigm. Creat Educ 6(02):224–239. https://doi.org/10.4236/ce.2015.62021
34. Bleakley A (2004) 'Your creativity or mine?': a typology of creativities in higher education and the value of a pluralistic approach. Teach Higher Educ 9(4):463–475. https://doi.org/10.1080/1356251042000252390
35. Ivcevic Z, Mayer JD (2009) Mapping dimensions of creativity in the life-space. Creat Res J 21(2–3):152–165. https://doi.org/10.1080/10400410902855259
36. Cohen L, Manion L, Morrison K (2013) Research methods in education. Routledge, New York
37. Krippendorff K (2004) Content analysis: an introduction to its methodology. Sage, Thousand Oaks

38. Maksić S, Pavlović J (2011) Educational researchers' personal explicit theories on creativity and its development: a qualitative study. High Ability Stud 22(2):219–231. https://doi.org/10.1080/13598139.2011.628850
39. American Heritage Dictionary of the English Language (2016) (5th Ed.). Houghton Mifflin Harcourt Publishing Company. http://www.thefreedictionary.com/word+cloud Accessed 15 June 2016

Paula Catarino is a Researcher at CMAT-UTAD and CIDTFF and has been an Associated Professor of Mathematics Department at the University of Trás-os-Montes e Alto Douro (UTAD) since 1985, in Vila Real, Portugal, where she teaches Linear Algebra. She is an Integrated Member of CMAT-UTAD (University of Minho's CMAT Polo) and a Collaborating Member of Lab-DCT/UTAD (Research Centre on Didactics and Technology in the Education of Trainers, CIDTFF, from University of Aveiro, and in its Lab-DCT/UTAD, the laboratory at UTAD). Her main research interests are related to algebra, more precisely with sequences of integers defined by recurrence relations. In addition, she is interested in the area of mathematical education, as well as in the Ethnomathematics research field. Recently, the creative and critical thinking research field in higher education is her other research interest. P. Catarino is a member of the Portuguese Engineers Society (OE)

Maria M. Nascimento has been an Assistant Professor at the University of Trás-Montes e Alto Douro (UTAD) since 1985, in Vila Real, Portugal, where she teaches Statistics and Operations Research. She is an Integrated Member as a Researcher of CIDTFF (Research Centre on Didactics and Technology in the Education of Trainers, CIDTFF, from University of Aveiro, and in its Lab-DCT/UTAD, the laboratory of CIDTFF at UTAD). Her main interests are teaching Statistics and its attitudinal and didactical research issues, as well as the Ethnomathematics research field. In addition, the critical thinking research field and its connections to the statistical thinking are her other research interests. Maria M. Nascimento is a Member of the Portuguese Engineers Society (OE) and of the Portuguese Society for Engineering Education (SPEE).

Eva Morais has been a Lecturer at University of Trás-os-Montes e Alto Douro since 2000, where she teaches courses in Statistics and Experimental Design. She is a Member at CMAT-UTAD (University of Minho's CMAT Polo), and her main research interests are related to the study of methods used to solve partial differential equations in financial pricing models, and she is also interested in the critical thinking in the higher education research field.

Paulo Vasco is a Researcher at CMAT-UTAD (University of Minho's CMAT Polo) and has been teaching at the University of Trás-os-Montes e Alto Douro (UTAD) in Vila Real, Portugal, since 2001, where currently he is an Assistant Professor, teaching mainly Linear Algebra at different courses including several of engineering. His main research interests are related to numerical semigroups, several number sequences defined by recurrence as well as Mathematics Education. Recently, he began to studying the critical thinking and creativity research fields and its connections to the area of mathematics.

Helena Campos is a Researcher at CIDTFF (Research Centre on Didactics and Technology in the Education of Trainers, CIDTFF, from University of Aveiro, and in its Lab-DCT/UTAD, the laboratory at UTAD) and has been an Assistant Professor at the University of Trás-Montes e Alto Douro (UTAD) since 2008, in Vila Real, Portugal, where she teaches basic Geometry and Didactics of Geometry. Her main research interests are in the area of mathematical education, more precisely in teaching Geometry and its attitudinal and didactical issues. In addition, she is interested in the area of algebra, namely with sequences of integers defined by recurrence relations.

Helena Silva is an Associate Professor at the University of Trás-Montes e Alto Douro (UTAD), in Vila Real, Portugal, since 2001. Her main research and teaching interests are related to teaching methodologies and teachers' professional development, with emphasis on cooperative learning, formative assessment, and communities of practice. She works with educators who seek to create classrooms that are more effective in academics and social skills. And recently, the critical thinking research field was discovered in its connections to professional development. She is also an educational author. She is a Researcher at CIIE—Centre for Research and Educational Intervention, University of Psychology and Educational Sciences, University of Porto, Porto, Portugal. She is also an educational author.

Rita Payan Carreira is a Researcher at CECAV—Centre for Animal and Veterinary Research—and has been an Assistant Professor at the Zootechnia Department, at University of Trás-Montes e Alto Douro (UTAD), in Vila Real, Portugal. Her research interests also cover educational issues, particularly related to active learning, and the development of critical thinking, inter- and intra-professional communication, and decision-making skills.

Maria João Pinto Monteiro is a Coordinating Professor at the University of Trás-Montes e Alto Douro (UTAD), in Vila Real, Portugal, since 2009, teaching mainly Nursing and Pedagogy in Health. She is a Researcher at CINTESIS—Center for Health Technologies and Services Research. Recently, the critical thinking research field became a research interest in its connections to professional development.

Printed by Printforce, the Netherlands